Structural Mechanics

A fully revised edition
of **Structural Mechanics**
by W Morgan and D T Williams

Frank Durka

Senior Lecturer in Structures
Department of Surveying
Liverpool Polytechnic

Longman Scientific & Technical

Longman Scientific & Technical
an imprint of
Longman Group Limited
Longman House, Burnt Mill, Harlow
Essex CM20 2JE, England
Associated companies throughout the world

First published in Great Britain by Pitman Publishing Limited 1980
Reprinted with corrections 1983
Reprinted 1984, 1985
Reprinted by Longman Scientific & Technical 1986

ISBN 0-582-98881-0

Printed in Great Britain at The Bath Press, Avon

Preface

The introduction of the metric system and other develop-
ments as the establishment of the Polytechnics and the setting
up of the Business and Technician Education Council have
created a need for this new edition.

Taking advantage of the opportunity presented by this
need, the text has been extensively revised to make it suitable
for students taking degree and diploma courses, particularly
in Architecture, Building and Surveying, but also for those in
the early stages of their studies in Civil and Structural En-
gineering. The adoption of SI units called for changes in the
treatment of the concepts of forces and stresses. This, in turn,
lead to some rearrangement of the material which now covers
the syllabus of the BTEC standard unit in Structural Mechanics
III (U76/397).

All the data based on British Standards have been updated
and, where appropriate, new material has been included, e.g.
welded and HSFG bolted connections, reference to limit
state design. (Extracts from British Standards appear by
permission of the British Standards Institution, 2 Park Street,
London, W1A 2BS, from whom complete copies may be
obtained.) The British Constructional Steelwork Association
(BCSA) and the Constructional Steel Research and Development Organisation (CONSTRADO) have also given permission to reproduce the tables shown on pages 156–161 and
205–207 from the 1978 edition of the *Structural Steelwork
Handbook* published jointly by BCSA/CONSTRADO.

The preparation of this fully revised edition of the work of
the late W. Morgan and D. T. Williams has not been confined
to a single person, as the title page may suggest. In this work
I have been helped by many, either directly or indirectly, and
I am grateful to them all. In particular my thanks go to John
Cushion of Pitman's in the first place for asking me to do it,
thus introducing me to a new, greatly gratifying although at
times extremely exacting experience, and especially for hi

peculiarly patient firmness in guiding me through the various stages of the project; to H. E. Walker for his thoroughness in reading the manuscript and for his pertinent comments; to my colleagues, Stan Houghton for his valued advice and always readily given assistance, and Alan Wilson for his enthusiastic interest and insistent encouragement.

None of this, however, would have been possible without the generous co-operation of my wife. I am, therefore, extremely grateful to her for putting up with the inconvenience caused by the whole exercise, and not least for all those innumerable mugs of coffee she so uncannily provided at the right moments when, in spite of the willingness of the spirit, the flesh grew weak.

I hope that many students will find this third edition as helpful in their studies as the first edition was to me and the second to most of my students. Then all the effort put into its preparation will have been well worthwhile.

Frank Durka
Liverpool 1980

Contents

Important Tables

1 Introduction

THE STRUCTURAL DESIGN

The fundamental purpose of a structure is to transmit loads from the point of application to the point of support and, ultimately, through the foundations to the ground. That purpose must be fulfilled within the constraints imposed by the client's brief, which will inevitably insist on low initial costs and low maintenance costs and may vaguely stipulate the functional needs of the project. Consequently the process of structural design begins with the appraisal of the client's requirements through collaboration with other members of the project design team—the architect, services engineer, quantity surveyor and, in some cases, the builder may be consulted at this stage as well.

The structural design itself is a combination of art and science in that it consists in the creation of a structural form which will accommodate the often conflicting aspects of cost, function, services, not forgetting aesthetics, and be capable of being quantified to produce dimensioned details for the purpose of erection. There are, therefore, two distinct stages in the process. In the first, the structural engineer draws on his experience, intuition and knowledge to make an imaginative choice of a preliminary scheme in terms of layout, materials and erection methods. In the second stage, the chosen scheme is subjected to detailed analysis based on the principles of structural mechanics. The resulting scheme must be consistent with the engineer's basic aims to provide a structure which satisfies the criteria of SAFETY and SERVICE-ABILITY at reasonable cost.

A safe structure may not be easy to define but it must, at least, not collapse under an applied load. The standards of safety have changed considerably since the Code of Hammurabi (c.2000 B.C.) declared the life of the builder forfeit should the house he built collapse and kill the owner. It was not until the 19th century, however, that the concept of a **factor of safety** was first formulated in terms of the ratio of ultimate or, in certain cases, yield stress to the working stress. This is explained in Chapter 6. Some decades later an alternative approach was introduced in terms of the **load factor** based on the ratio of loads rather than stresses, i.e. load factor = ultimate-load/working-load. Using factors of safety ensures a satisfactory performance under working loads, but only presupposes a reasonable margin against failure. Load factors, on the other hand, ensure a definite factor against failure and assume a satisfactory performance under working loads.

Both factors operate on the implicit assumption that the determined values both for the loading which the structure is expected to carry, and for the strength of the materials of which the structure is made, remain constant, thus inferring a guarantee of absolute safety. Whilst this may be satisfactory in certain cases, it is generally recognized that a more realistic measure of the safety of a structure can be achieved through an assessment of the **probability** of its failure. In this approach, partial safety factors are used, separating the probability of failure due to overloading from that due to variability of strength of the materials.

Whereas safety deals with the structure's ability to carry its loads without unwarranted risks to human life and limb, the requirements of serviceability of the structure insist on its fitness for use without excessive deflections, disturbing vibrations, noticeable cracking and other local failures which may demand costly remedial work.

MATHEMATICAL MODELS

To ensure the safety of the structure it is necessary to provide it with sufficient STRENGTH and, to safeguard its serviceability, it must have adequate STABILITY. In order to assess the sufficiency of its strength and the adequacy of its stability, the structure needs to be presented in quantitative terms before it can be analysed and, since it does not yet exist, its behaviour has to be simulated. A scaled-down version of the proposed structure may seem the obvious choice here, but now even fairly complex structures can be analysed by means of mathematical models.

Consider, for example, the behaviour of the steel spring shown in Fig. 1.1. As the load P is increased, the extension x in the length of the spring will also increase. From several pairs of measurements of P and x it will be found that x is proportional to P. Therefore, the relationship between the

Original length of spring

Extension x

Load P

Figure 1.1

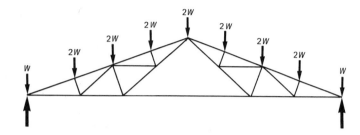

Figure 1.2 Steel roof truss

Figure 1.3 Idealised representation of truss shown in Fig. 1.2

two quantities can be formulated by means of the equation $P = k \times x$, where k is a constant representing, in this case, the stiffness of the spring. This equation serves as the mathematical model for the behaviour of the loaded spring since it allows the extension in the length of the spring to be determined for any given load.

The behaviour of a structure is clearly more complex than that of a steel spring. It calls, therefore, for more complex models. In fact, even a simple structure may contain too many unknown factors to be analysed completely. In such cases it becomes necessary to replace the real structure by a simplified, idealized version of it.

An example of this simplification can be shown in the case of a steel roof truss. Fig. 1.2 represents the real structure. Note that the roof loads are applied to the truss at the joints; the members of the truss are made up of standard structural steel sections and are welded together. Fig. 1.3 shows an idealized truss. Here the roof loads are indicated by arrows directed vertically downwards, and the truss members are shown as simple bars with pinned connections.

These two structures differ in several aspects, the most significant being the substitution of pinned connections for the welded ones. A welded connection gives some degree of rigidity which complicates the behaviour of the truss members. A pinned connection, on the other hand, allows the members free movement relative to each other, thus simplifying their analysis. This idealization is made on the assumption that the truss would function satisfactorily if the connections were in fact pinned, and that the difference in the behaviour of the members of the real and the idealized trusses is sufficiently small. This has been shown by experiment and experience to be the case for the ordinary range of trusses.

It is, of course, of vital importance that the structural engineer keeps these assumptions within strict limits in the knowledge that each assumption tends to reduce the accuracy of the subsequent analysis. The engineer must be aware, for example, that the application of the model for the behaviour of the steel spring, discussed earlier, is restricted by the fact that the stiffness of the spring remains constant only within certain range of values for P. The validity of the model, after all, depends on how closely it represents the real behaviour. The engineer is required, therefore, to make grave and far-reaching choices which must be founded on sound experience, supported by an intuitive understanding and fostered by a broad knowledge of structural behaviour. In the search for the appropriate solution, the engineer may find assistance in the Codes of Practice, which are based on the collective opinion and corporate judgement of experienced designers. They are intended to provide a set of guidelines within which design decisions can be formulated and the analyses carried out in accordance with the principles of Structural Mechanics.

FORCES

The term mechanics, according to a dictionary definition, refers to "that branch of applied science which deals with the action of forces in producing motion or equilibrium". In the context of Structural Mechanics, forces represent the loads that the structure is expected to carry. These are generally classified into three groups: the dead loads, the imposed (live) loads, and wind loads.

Dead loads comprise the permanent loads due to the static weight of the structure itself, the cladding, floor finishes, and any other fixtures which form the fabric of the building.

The **imposed (live) loads** are those produced by the intended occupancy of the building, i.e. loads due to the weight of plant and equipment, furniture, the people who use the building, etc. They also include snow loads, impact and dynamic loads arising from machinery, cranes and other plant, and such irregular loads as those caused by earthquakes, explosions and changes in temperature.

Wind loads are classified separately owing to their transitory nature and the complexity of their effects.

The structure may be subjected to the action of almost any combination of these loads and, in order to produce equilibrium, it must provide an equal and opposite **reaction.** The necessary reaction is generated by the STRESS caused by the action of the loads within the material, and by the ensuing STRAIN in the elements of the structure.

The two concepts (stress and strain) are discussed in detail in Chapter 6. They are the direct outcome of the action of FORCES and the DEFORMATIONS they produce.

The definition of a force is generally derived from Newton's first law of motion as that which changes or tends to change an object's uniform motion in a straight line or, as is more appropriate to structural mechanics, its state of rest.

Most of the loads supported by a structure exert forces on it by virtue of their mass which is subjected to the gravitational pull of the earth. This pull, as was demonstrated in the case of the steel spring in Fig. 1.1, acts downwards, thus indicating that the force has a direction as well as a magnitude and is, therefore, a vector quantity.

As such, a force can be represented by a straight line of a given length indicating the magnitude of the force. The direction of this straight line is drawn parallel to the line of action of the force and denoted by an arrow. Such a line is called a **vector.**

The SI unit of force is the newton, denoted by the capital letter N, and is that force which, when applied to an object having a mass of one kilogramme, gives it an acceleration of one metre per second per second, i.e. $N = kg \times m/s^2$.

Attention must be drawn here to the ambiguity that exists in the use of the term weight. In this textbook, weight is considered as the force exerted on a body by the gravitational pull of the earth and is, therefore, stated in force units (newtons). However, in accordance with the Weights and Measures Act 1963, the weights of structural materials are specified in mass units (kilogrammes). Hence, in order to calculate the forces due to dead loads, the kilogrammes have to be converted to newtons by multiplying them by the gravitational acceleration. The value of this acceleration is $9.81 \, m/s^2$, but it is generally accepted that, to simplify the conversion for the purpose of structural calculations, the figure may be taken to be $10 \, m/s^2$. For example the gravitational force due to a 100 kg weight is

$$100 \, kg \times 10 \, m/s^2 = 1000 \, N \quad or \quad 1 \, kN$$

To counteract the action of the gravitational (and other) forces, the structure must provide an equal and opposite reaction in order to remain in a state of rest, i.e. in equilibrium. The action of the forces may manifest itself in any of the following ways:

Tension   The forces acting on the object act away from each other; the object (string in Fig. 1.4) is being stretched.

Reaction = 250N
(neglecting weight
of string)

Tension in string = 250N

Pull downwards = 250N

Figure 1.4

Figure 1.5

Compression The forces acting on the object act towards each other; the object (brick pillar in Fig. 1.5) is being squeezed.

Shear The forces acting on the object cause parts of the object to slide in relation to each other (Fig. 1.6).

Bending The forces acting on the object produce a bending effect as shown in Fig. 1.7. Note that this action also causes sliding (shear).

Torsion The forces acting on the object cause it to be twisted (Fig. 1.8).

Figure 1.6

Figure 1.7

Figure 1.8

Force, however, cannot be observed directly. It is recognized and, therefore, measured by the deformations it causes. The deformation which is of particular interest to the structural engineer is that resulting from bending, i.e. deflection.

The control of deflection, for example, is of great importance from the point of view of the stability of the structure. It must, therefore, be quantified (calculated). The study of the deflected forms of structural elements or frameworks, however, provides a useful basis for the appreciation of structural behaviour and it is advisable to acquire this helpful ability.

A number of simple examples is illustrated in Figs 1.9–1.15.

Figure 1.9 Straight cantilever

Figure 1.10 Propped cantilever

Figure 1.11 Encastré beam

Figure 1.12 Simply supported beam

Figure 1.13 Continuous beam

Figure 1.14 Three-hinged arch

Figure 1.15 Portal frame

2 Concurrent Coplanar Forces

All the problems in this chapter are concerned with forces which

   i) meet at a point, i.e. they are concurrent, and
   ii) act in one plane, i.e. they are coplanar.

The chapter deals with the graphical methods of resolving such forces into convenient components, and then determining their resultant effect, for the purpose of establishing the equilibrium of the system of forces.

TRIANGLE OF FORCES

AB in Fig. 2.1 represents a horizontal wooden beam containing a number of hooks. By means of string, a small ring, and two spring balances a weight of 35 N is suspended as shown. It is assumed in this example that the readings on the balances are 20 N and 25 N respectively. The ring is in equilibrium (at rest) under the action of three forces, i.e. the vertical pull downwards of the weight and the pulls exerted by the two strings. The condition can be represented on drawing paper by Fig. 2.2 which is called the **free-body diagram** with respect to point O.

The three lines are called the **lines of action** of the forces and the arrows represent the **direction** in which the forces are acting on the ring. In order to plot on paper the lines of action of the forces, the angles between the strings can be measured by means of a circular protractor. An alternative and simpler method of transferring to paper the lines of action is to hold a sheet of paper or cardboard behind the strings and, by means of a pin, to prick two or three points along each string. Pencil lines can then be drawn to connect these points and thus to fix on paper the free-body diagram.

Figure 2.1

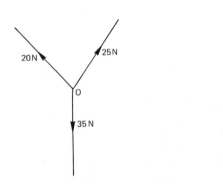

Figure 2.2   Free-body diagram for point O

Figure 2.3 Force diagram

(a)

Figure 2.4

(b)   (c)

Figure 2.5

(a)   (b)

Figure 2.6 (a) Free-body diagram for point O; (b) Force diagram

The magnitudes of the forces can be represented on paper by vectors, i.e. lines drawn to scale. For example, if it is decided to let 1 mm represent 1 N then the force of 35 N will be represented by a line 35 mm long, the force of 25 N by a line 25 mm long, etc.

Now, if the free-body diagram is drawn on paper to represent the lines of action and directions of the forces, and three lines are drawn, as shown in Fig. 2.3, parallel to these forces to represent to scale their magnitudes, it will be found that a triangle will be formed. In an actual experiment the result may not be quite as accurate as indicated in Fig. 2.3 because of possible lack of sensitivity of the spring balances and the difficulty of transferring accurately to paper the free-body diagram. The fact is, however, that,

if three forces meeting at a point are in equilibrium, they may be represented in magnitude and direction by the three sides of a triangle drawn to scale.

This is the **principle** (or **law**) **of the triangle of forces.**

By placing the spring balances on different hooks, e.g. as in Fig. 2.4, further illustrations of the principle of the triangle of forces can be obtained.

Action and reaction are equal and opposite, yet in Fig. 2.4 it appears that 85 N are required to support a downward force of 35 N. This is because the strings are inclined in opposite directions and react on each other as well as on the force of 35 N. For example, the string which has a tension of 40 N is pulling towards the left in addition to pulling up. This pull to the left is resisted by the other string pulling to the right, so that only a part of the force in each string is effective in holding up the weight. This will be demonstrated more formally on page 14.

It is advisable to carry out a few experiments similar to those of Figs. 2.1 and 2.4 in order to verify the law of the triangle of forces. The law can then be used to determine unknown forces and reactions.

Example 2.1

Determine the tensions in the two strings L and R of Fig. 2.5.

Solution Point O is in equilibrium under the action of three forces as shown in the free-body diagram of Fig. 2.6, therefore the triangle of forces law must apply.

Draw a line *ab* parallel to the force of 100 N to a scale of so many newtons to the millimetre. From *b* draw a line parallel to string L. From *a* draw a line parallel to string R. The two lines intersect at point *c*. The length of line *bc* gives the tension in string L (i.e. 81.6 N) and the length *ca* gives

Figure 2.7

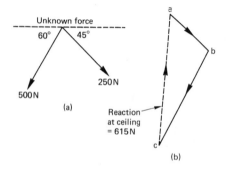

Figure 2.8 (*a*) Free-body diagram; (*b*) Force diagram

Figure 2.9

the tension in string R (i.e. 111.5 N). (A student with a knowledge of trigonometry can obtain the magnitudes of the forces by calculation.)

Example 2.2

Two ropes are attached to a hook in the ceiling and a man pulls on each rope as indicated in Fig. 2.7. Determine the direction and magnitude of the reaction at the ceiling (neglect the weights of the ropes).

Solution Since the hook is in equilibrium under the action of the forces in the ropes and the reaction at the ceiling, the triangle of forces principle applies. Draw *ab* parallel to the force of 250 N to a suitable scale (Fig. 2.8). From *b* draw *bc* parallel to the other rope to represent 500 N. The closing line *ca* of the force triangle represents to scale the reaction at the ceiling and is nearly 615 N, acting at approximately 7° to the vertical. This force can be called the **equilibrant** of the two forces in the ropes, since the ceiling must supply a force of 615 N in order to maintain equilibrium.

In the last problem, the reaction at the ceiling due to the pulls in the two ropes equalled 615 N acting as shown in Fig. 2.9. If we substitute one rope for the two ropes and pull with a force of 615 N in the direction indicated, the effect on the ceiling will be exactly as before, i.e. the result of the two forces is equivalent to one force of 615 N. This force is, therefore, called the **resultant.** Note also that this resultant is equal in value to the equilibrant and acts in the same straight line but in the opposite direction. The resultant of a given number of forces is therefore the single force which has the same effect on the equilibrium of the body as the combined effects of the given forces.

PARALLELOGRAM OF FORCES

The law of the parallelogram of forces is in essentials the same as the law of the triangle of forces. Any problem which can be solved by the parallelogram of forces law can be solved by the triangle of forces law, although it may sometimes be slightly more convenient to use the former method.

Example 2.3

Referring to Example 2.2 determine the resultant of the pulls in the two ropes by using the parallelogram of forces.

Figure 2.10

Figure 2.11

Figure 2.12

Figure 2.13

Solution Measure a distance *oa* along line OA in Fig. 2.10 to represent 500 N and measure a distance *ob* along line OB to represent 250 N. From *a* draw a line parallel to OB and from *b* draw a line parallel to OA thus forming a parallelogram. The length of line *oc*, which is the diagonal of the parallelogram (from the point where the two forces meet), gives the value of the resultant of the two forces (615 N) and also its direction. Note that the triangle *obc* is identical with the triangle of forces in Fig. 2.8.

Example 2.4

Two ropes pull on an eye bolt as indicated in Fig. 2.11. Determine the resultant pull on the bolt.

Solution The solution is indicated in Fig. 2.12.
Note that when the parallelogram of forces law is used to determine the resultant of two forces, the two forces must be drawn so that the arrows representing their directions either both point towards the meeting point or both point away from it. In other words, a thrust and a pull must be converted either into two thrusts or into two pulls.

Example 2.5

Determine the resultant force on the peg due to the thrust of 2.5 kN and the pull of 1.5 kN (Fig. 2.13).

Solution At first glance this problem might appear identical with that given in Fig. 2.11. Obviously, however, the resultant of these two forces must be different from the resultant of the two forces in Fig. 2.11.
The resultant is found either as shown in Fig. 2.14 or as shown in Fig. 2.15. Note that the effect on the peg is the same whether it is pushed or pulled with a force of 1.8 kN.

Figure 2.14   **Figure 2.15**

Figure 2.16

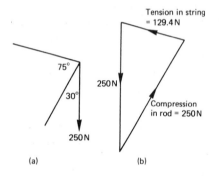

Figure 2.17 (*a*) Free-body diagram for point O; (*b*) Force diagram

Figure 2.18

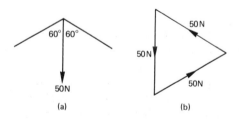

Figure 2.19 (*a*) Free-body diagram; (*b*) Force diagram

Example 2.6

A rigid rod is hinged to a vertical support and held at 60° to the horizontal by means of a string when a weight of 250 N is suspended as shown in Fig. 2.16. Determine the tension in the string and the compression in the rod, ignoring the weight of the rod.

Solution Point O is in equilibrium under the action of three forces which meet at the point, i.e. the weight of 250 N, the tension in the string and the compression in the rod. The solution is given in Fig. 2.17. Note that a tension member is called a **tie** and a compression member is called a **strut.**

Example 2.7

Fig. 2.18 shows two rods of wood connected by a hinge at the top and supported by two concrete blocks. Determine the compressions in the rods ignoring their own weight.

Solution Considering the equilibrium at the hinge, the free body and force diagrams are shown in Fig. 2.19. The compression in each rod is 50 N. If the angle of slope of the members is made greater, e.g. in Fig. 2.20, the compressions in the members are less than in Fig. 2.19.

RECTANGULAR COMPONENTS

The apparatus of Fig. 2.18 is shown in elevation in Fig. 2.21. If a gradually increasing pull is applied to the spring balance, the vertical force necessary to cause lifting of the blocks can be recorded. Assume this downward force is 75 N when the angles are as given and when the blocks just begin to lift. The compression acting along each member is 75 N as obtained by the force diagram. Now consider what happens at the bottom end of one compression member. The member is pushing on the block with a force of 75 N (Fig. 2.22). Since this force is inclined it has a twofold effect. It is tending to cause crushing of the concrete as well as to cause overturning. If the force were vertical (and axial), there would be crushing effect only and no overturning effect. If the force were horizontal, there would be overturning effect and no crushing effect.

It is sometimes convenient to **resolve** such a force into its **horizontal component** and its **vertical component.**

It was demonstrated on page 11 that the resultant of two forces meeting at a point is given by the diagonal of a

Figure 2.20

Figure 2.23

Figure 2.21   **Figure 2.22**

parallelogram. By a reverse process, one force can be replaced by (or resolved into) two forces. By constructing a rectangle (parallelogram) (Fig. 2.23) with the force in the rod (75 N) as the diagonal, the horizontal and vertical components are given by the two sides respectively of the rectangle. The components can be found either by drawing to scale or by calculation.

Example 2.8

Replace the tie rope shown in Fig. 2.24 by two ropes, one vertical and one horizontal which together will have the same effect on the eye bolt.

Figure 2.24

Solution By constructing a rectangle with the force of 20 kN, drawn to scale, as the diagonal (Fig. 2.25) the vertical component (and therefore the force in the vertical rope) is found to be 12.9 kN. The horizontal component is 15.3 kN.

In the above problem, the components are called **rectangular components** because the angle between them is 90°. When a force has to be split into two components it is usually the rectangular components which are required. It is possible, however, to resolve a force into two components with any given angle between them.

Figure 2.25

Example 2.9

Replace the rope X which is pulled with a force of 500 N by two ropes L and R as indicated in Fig. 2.26.

Solution Measure a length along line AB to represent 500 N and assume this line to be the diagonal of a parallelogram by drawing lines from B parallel to ropes L and R respectively. The forces in the two ropes which together have the same effect on the bolt as the single rope can now be scaled off, i.e.

Figure 2.26

Figure 2.27

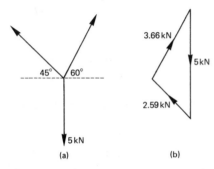

Figure 2.28 (*a*) Free-body diagram;
(*b*) Force diagram

Figure 2.29

Figure 2.30

Force in rope L = 450 N (length AC)

Force in rope R = 365 N (length AD)

These two forces can be said to be components of the original force of 500 N.

Example 2.10

Determine the tensions in the two chains X and Y of Fig. 2.27. Then determine the horizontal and vertical components of these tensions. In addition, determine the horizontal and vertical components of the reactions at A and B. Neglect the weight of the chains.

Solution The free-body and force diagrams for the point are given in Fig. 2.28. The tensions in the chains are respectively 2.59 kN and 3.66 kN. The horizontal and vertical components of the tensions in the chains are given in Fig. 2.29.

Note that the sum of the two vertical components is equal to the value of the suspended load of 5 kN (action and reaction are equal and opposite). Note also that the horizontal component of the force in chain X is equal to the horizontal component of the force in chain Y and acts in the opposite direction (action and reaction are equal and opposite).

Now consider the reactions at points A and B which equal the tensions in the chains (Fig. 2.30). Again, neglecting the weights of the chains, the sum of the vertical components of the reactions is equal to 5 kN and that the horizontal components are equal and opposite.

Referring back to Fig. 2.21 on page 13, it was stated that the concrete blocks were on the point of overturning when the compression acting along each member was 75 N. The horizontal component of this force is 65 N, and this component can be prevented from having an overturning effect by connecting the two inclined members by a horizontal tie member (Fig. 2.31). It will now be impossible to cause overturning of the concrete blocks by pulling downwards at the apex. The tension in the tie member will be equal to the horizontal component (65 N) of the thrust in the inclined member.

Another way of finding the tension in the tie is by applying the triangle of forces law to the joint between the inclined and horizontal members (Fig. 2.32). Neglecting the weights of the members, the vertical downward force of 75 N is balanced by the reactions of the blocks (37.5 N each block).

Figure 2.31

Figure 2.32

Figure 2.33 (*a*) Free-body diagram; (*b*) Force diagram

Figure 2.34

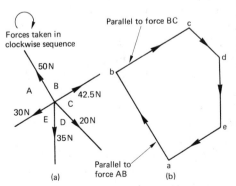

Figure 2.35 (*a*) Free-body diagram for meeting point of forces; (*b*) Force polygon

Consider the equilibrium of point O (Fig. 2.33). Note that, when three forces meeting at a point are in equilibrium, the arrows in the force diagram follow one another around in the same direction. The strut is pushing towards point O and the tie is pulling away from the same point.

POLYGON OF FORCES

The same apparatus used in Fig. 2.1 can be employed to demonstrate the principle of the polygon of forces.

Referring to Fig. 2.34, the small ring is at rest (i.e. in equilibrium) as the result of the downward pull of the 35 N force and the pulls in the four strings. The forces in the strings are given by the readings on the spring balances and these forces are given in Fig. 2.34 for one particular arrangement of the strings.

The lines of action of the forces can be transferred to a piece of paper or cardboard (in the manner described on page 8) to give the free-body diagram (Fig. 2.35).

It will be noted that the *spaces* between the forces in the free-body diagram have been lettered A, B, C, etc., and each force is therefore denoted by two letters, e.g. force AB (which separates space A from space B), force BC, force CD, etc. (If the forces are taken in anti-clockwise sequence they will be called force DC, force CB, force BA, etc., but it is more usual to take the forces in clockwise sequence.) This method of notation was devised by an engineer (R. H. Bow) about 1870 and is known as **Bow's Notation.** It is not essential to use a notation for simple problems but some form of notation is indispensable when a large number of forces are involved, such as in roof trusses (see Chapter 5).

Having plotted the lines of action of the forces in the free-body diagram, the next step is to choose a scale of so many newtons to the millimetre in order to plot the force diagram. Starting with the force AB, a line *ab* is drawn parallel to it, the length *ab* representing to scale the magnitude of the force (50 N). From *b* a line *bc* is drawn in the direction indicated by the arrow on force BC (42.5 N).

The other forces are drawn in the same manner and, if the experiment has been performed accurately, it will be found that the last line drawn parallel to force EA (30 N) will finish at *a*, where line *ea* represents to scale the force EA.

It is not necessary to start the force diagram with force AB. Any one of the forces can be chosen as the starting force.

Further experiments can be performed by connecting the strings to the other hooks in the framework, and it will be found in every case (apart from small experimental errors)

Figure 2.36

(a)

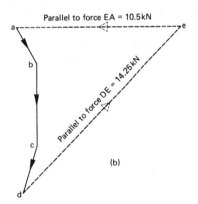

(b)

Figure 2.37 (a) Free-body diagram; (b) Force polygon

that, when a number of forces are in equilibrium, the force diagram will form a polygon, the sides of which represent to scale the magnitudes and directions of the forces.

Study carefully the force diagram or polygon in Fig. 2.35 and note that the arrows "chase each other" in the same sense around the diagram, i.e. the tip of each arrow is pointing to the tail of the arrow in front of it. This will always be true for force diagrams when the forces are in equilibrium, and use can be made of this fact to discover the direction of unknown forces, as in the following example.

Example 2.11

A rod (the weight of which is negligible) is hinged to a support at S and is supported by a tie (Fig. 2.36). From the point O three ropes are pulling with the forces indicated. Determine the forces in the rod and tie.

Solution The point O is in equilibrium as the result of the action of five forces, i.e. the three given forces and the forces in the rod and tie.

The free-body diagram for point O is given in Fig. 2.37.

It is an advantage to start Bow's Notation in such a manner that the two unknown forces are last. First, draw a line *ab* parallel to force AB to represent to scale 2.5 kN, then draw lines *bc* and *cd* to represent the other known forces. From *d* draw a line parallel to force DE. Since point O is in equilibrium the force diagram must form a closed polygon. Therefore from *a* draw a line parallel to force EA to intersect the line drawn parallel to force DE. The intersection point of these two lines is point *e*.

The force in the rod is given by the length of line *de* and is 14.3 kN approximately. Similarly, the force in the tie is given by the length of line *ea* and is 10.5 kN approximately. The arrows on lines *de* and *ea* must follow the general direction; therefore the tie is pulling away from point O and the rod (which is a strut) is pushing towards point O.

Note that the forces in the rod and tie have been stated to be 14.3 kN and 10.5 kN approximately. The forces can be obtained by calculation and may be slightly different from the values given, since calculation is more accurate than drawing. Graphical methods are, however, sufficiently accurate for most structural work and should be used when they are quicker and more easily applied than calculation methods.

The resultant of two forces meeting at a point can be found by the triangle or parallelogram of forces as described earlier. The resultant of any number of forces can be determined by the polygon of forces.

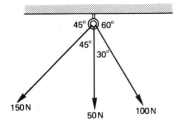

Figure 2.38

Example 2.12

Three ropes pull on a bolt as indicated in Fig. 2.38. Determine the resultant pull.

Solution The force polygon is constructed from *a* to *d* by drawing lines to scale parallel to the three given forces (Fig. 2.39). The closing line *da* of the polygon gives the magnitude and direction of the reaction (equilibrant) supplied by the support. The resultant of the pulls in the three ropes is, of course, equal to the equilibrant and is 249.1 N pulling downwards at an angle of approximately 13° to the vertical.

Figure 2.39 (*a*) Free-body diagram; (*b*) Force polygon

SUMMARY

Triangle of Forces. If three forces, the lines of action of which meet at one point, are in equilibrium, they can be represented in magnitude and direction by the sides of a triangle if these sides are drawn parallel to the forces.

Parallelogram of Forces. If two forces meeting at a point are represented in magnitude and direction by the two sides of a parallelogram, their resultant is represented in magnitude and direction by the diagonal of the parallelogram which passes through the point where the two forces meet.

Polygon of Forces. If any number of forces acting at a point are in equilibrium, they can be represented in magnitude and direction by the sides of a closed polygon taken in order.

Components of Forces. A given force can be replaced by any two forces (components) which meet at the point of application of the given force. When the angle between the two forces is a right angle the components are called rectangular components.

Equilibrant. The equilibrant of a given number of forces which are acting on a body is the single force which keeps the other forces in equilibrium. Any one of the forces can be considered as being the equilibrant of the remainder of the forces.

Resultant. The resultant of a given number of forces is the single force which, when substituted for the given forces, has the same effect on the state of equilibrium of the body.

EXERCISE 2

1 Determine the tensions in the two chains L and R in the case of Fig. 2.40 (*a*), (*b*) and (*c*).

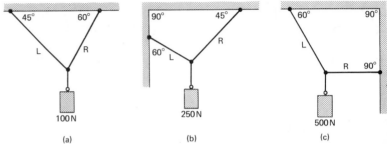

Figure 2.40

(a)   (b)   (c)

2 The angles marked A in Fig. 2.41 are equal. Determine the minimum value of this angle if the tension in each rope must not exceed 1 kN.

3 The string marked L in Fig. 2.42 passes over a smooth pulley. Determine the tensions in the two strings L and R by considering the equilibrium of point (1) then determine the reaction at the pulley by considering the equilibrium of point (2).

4 The string marked L in Fig. 2.43 passes over a smooth pulley and the system is in equilibrium. Determine the tensions in the strings L and R, then determine the magnitude of the weight W and the tension in the string M. (Consider first the equilibrium of point (1).)

5 Two struts thrust on a wall as indicated in Fig. 2.44. Determine the magnitude and direction of the resultant thrust.

6 The top of a pole resists the pulls from two wires as indicated in Fig. 2.45. Determine the magnitude and direction of the resultant pull.

7 A bolt resists the pull from three wires as indicated in Fig. 2.46. Determine the resultant of the two pulls of 5 kN and 2.5 kN then combine this resultant with the remaining pull of 4 kN to determine the resultant pull on the bolt.

8 Referring to Fig. 2.47, determine the value of the angle A so that the resultant force on the wall is vertical. What is the value of the resultant?

9 The top of a pole sustains a pull and a thrust as shown in Fig. 2.48. Determine the resultant force.

10 Two men pull on ropes attached to a peg as shown in Fig. 2.49. Determine the resultant pull on the peg.

11 In trying to move a block of stone resting on the ground, one man pushes and another man pulls as shown in plan in Fig. 2.50. Determine the resultant of the two forces.

12 Determine the resultant pull on the bolt in Fig. 2.51 and the reactions at the pulleys.

Figure 2.41

Figure 2.42

Figure 2.43

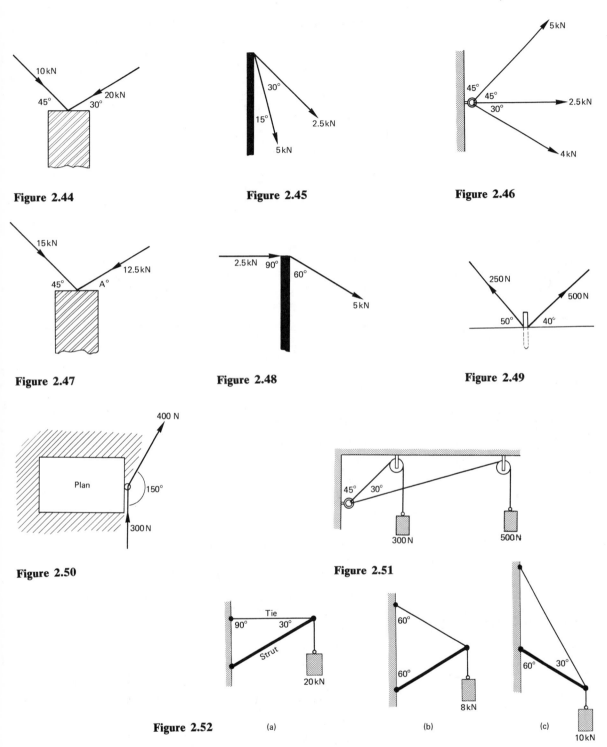

Figure 2.44

Figure 2.45

Figure 2.46

Figure 2.47

Figure 2.48

Figure 2.49

Figure 2.50

Figure 2.51

Figure 2.52   (a)   (b)   (c)

13 Determine the forces in the tie and strut in the case of Fig. 2.52 (a), (b) and (c) neglecting the weights of the members. All joints are hinged.

Figure 2.53

Figure 2.54

Figure 2.55

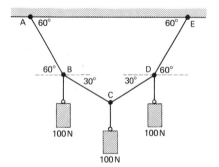

Figure 2.57

14 Referring to Fig. 2.53 a weight of 5 kN is suspended from A by a flexible cable. The rope C is parallel to the member AB. Determine the force in the cable, the angle x and the forces in the strut and tie. Ignore the weights of the members.

15 Determine the forces in the two struts of Fig. 2.54 then determine the vertical and horizontal thrusts at each support.

16 Fig. 2.55 shows a simple roof truss consisting of two struts and a tie. Consider the equilibrium of joint C to determine the forces in the two struts, then, using these values, apply the triangle of forces law to joints A and B respectively to determine the tension in the tie and the vertical reactions at the supports.

17 Determine the forces in the tie and strut of Fig. 2.56, then determine the vertical and horizontal components of the reactions at A and B.

18 A continuous string ABCDE has weights suspended from it as shown in Fig. 2.57. Determine the tensions in the portions AB and BC. In addition, determine the horizontal and vertical components of the reactions at A and E.

19 In Fig. 2.58 the masts are vertical and hinged at their bases. Determine the tension in the rope supporting the weight, the tension in the stays and the compression in the masts.

20 In Fig. 2.59 the rope supporting the weight passes over smooth pulleys at the tops of the masts. Determine the angle A that the masts make with the horizontal and determine the compression in the masts.

Figure 2.56

Figure 2.58

Figure 2.59

Figure 2.60

Figure 2.61

21 The block of stone of Fig. 2.60 will begin to slide when the horizontal force is 2.5 kN. What is the maximum value of the thrust P without sliding occurring?

22 Determine the force in the ropes A and B which will replace the rope C of Fig. 2.61.

23 Two posts support a weight of 1 kN as shown in Fig. 2.62. What horizontal force is each post exerting on the other?

24 A triangular framework of hinged rods is supported by two cables (Fig. 2.63). Determine the forces in the framework and in the cables, and determine the vertical components of the reactions at A and B. Neglect the weights of the members.

25 A framework consisting of five rods hinged at their joints is supported as shown in Fig. 2.64. Neglecting the weights of the members, determine the forces in the members of the framework and the reaction at A.

Figure 2.62

Figure 2.63

Figure 2.64

Figure 2.65

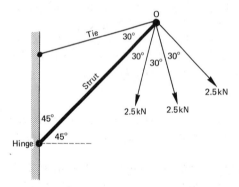

Figure 2.66

26 Fig. 2.65 (a) to (f) shows the free-body diagrams for systems of concurrent forces which are in equilibrium. Determine the magnitude and direction of the unknown force (or forces) marked X and Y.

27 Three ropes are attached to point O and are pulled with forces as indicated in Fig. 2.66. Determine the forces in the strut and tie.

3 Non-concurrent Coplanar Forces

Figure 3.1

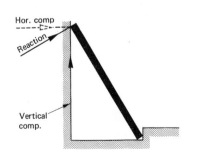

Figure 3.2

For the solution of the problems in this chapter it is necessary to know the following facts:

1 The whole weight of a body can be assumed to act at one point for the purpose of determining supporting reactions, etc., in a system of forces which are in equilibrium. This point is called the **centre of gravity** of the body. This subject will be discussed in detail in Chapter 8. For the present, it will be sufficient to remember that the centre of gravity (c.g.) of a rod or beam which is uniform in cross-section is at the centre of its length; the c.g. of a thin rectangular plate is at the intersection of the diagonals of the rectangle; the c.g. of a thin circular disc is at the centre of the circle; and the c.g. of a cube or rectangular solid is at the intersection of its diagonals.

2 If a body is supported by a perfectly smooth surface, the reaction of the surface upon the body acts at right angles to the surface (Fig. 3.1).

If the reaction were not at right angles to the surface, it would mean that there would be a component of the reaction acting along the wall surface as shown in Fig. 3.2. This shows there is frictional resistance of the wall opposing the downward trend of the ladder, and if the wall is smooth it can offer no frictional resistance.

3 If two forces (which are not parallel) do not meet at their points of contact with the body, their lines of action can be produced until they do meet.

4 If a body is in equilibrium under the action of three forces, and two of these forces are known to meet at a point, the line of action of the remaining force must pass through the same point.

Figure 3.3

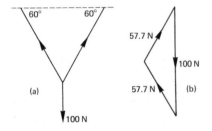

Figure 3.4 (*a*) Free-body diagram for point O of Fig. 3.3; (*b*) Force diagram

Figure 3.5 (*a*) Ladder resting against a smooth wall; (*b*) Free-body diagram for point O; (*c*) Force diagram

Example 3.1

A thin rectangular plate weighing 100 N is supported by two strings as indicated in Fig. 3.3. Determine the tensions in the strings.

Solution The whole weight of the plate can be assumed to act at its centre of gravity. The lines of action of the forces in the strings, when produced, meet the line of action of the weight of the plate at O. Consider the equilibrium of this point (Fig. 3.4).

The tension in string A (and B) is 57.7 N. The principle involved in the solution of the above problem can be demonstrated on the apparatus illustrated in Fig. 2.1.

Example 3.2

A ladder rests against a smooth wall and a man weighing 750 N stands on it as indicated in Fig. 3.5. Neglecting the weight of the ladder determine the reactions at the wall and at the ground.

Solution The ladder is in equilibrium as the result of the action of three forces: the weight of the man, the reaction at the top of the ladder, and the reaction at the foot. Since the wall is smooth the reaction at the top of the ladder is at right angles to the wall. Draw Fig. 3.5(*a*) to scale. The vertical line representing the weight of the man and the horizontal line representing the reaction at the top of the ladder meet at point O. The line of action of the reaction at the foot of the ladder must therefore pass through the same point. We now have three forces meeting at a point and the solution is shown in Figs. 3.5(*b*) and (*c*) which give the reaction at the top of the ladder as 125 N, and the reaction at the foot as 760 N.

Figure 3.6

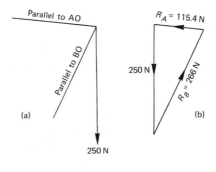

Figure 3.7 (*a*) Free-body diagram for point O; (*b*) Force diagram

Example 3.3

The uniform rod 2 m long shown in Fig. 3.6 weighs 250 N. Determine the reactions at A and B.

Solution The reaction at A is in line with the string. The weight of the rod can be assumed to act at its centre of gravity and to act vertically downwards. Draw the diagram to scale. Produce the line of action of the force of 250 N to meet at O the line of action of the reaction at A. The line of action of the reaction at B is given by joining B to O. Consider the equilibrium of point O as shown in Fig. 3.7. The reactions are 115.4 N at A and 266 N at B.

THE LINK POLYGON

Non-concurrent forces are forces whose lines of action do not all meet at one point. (A system in which the forces are all parallel to one another is a special example of non-concurrent forces.)

Fig. 3.8 shows a system of non-concurrent forces. Assume that the forces are acting on a block of stone, the plan view of which is a square of 0.8 m side, the forces being coplanar. If the resultant force on the stone is required, one method of solution based on work in the previous chapter is first to produce any two of the forces until they meet, and then to find the resultant of these two forces by the parallelogram of forces. This resultant can now be combined with one of the remaining forces and a new resultant found, and so on until the forces have been reduced to one resultant. This method, however, is cumbersome when the number of forces is large and a much neater solution is shown in Fig. 3.9.

Figure 3.8

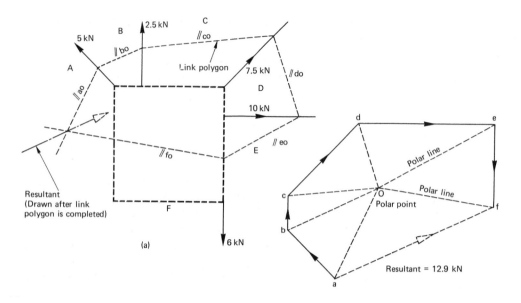

Figure 3.9 (a) Free-body diagram with link polygon; (b) Force polygon with polar diagram

The free-body diagram giving the lines of action of the forces is drawn accurately and the forces lettered with Bow's Notation. These forces are treated as if they all meet at one point and a force polygon is drawn, the sides of which are parallel to the given forces.

The closing line *fa* (direction from *f* to *a*) of the force polygon will give the magnitude and direction of the equilibrant of the five forces and the same line in the direction *a* to *f* will give the resultant. If the five forces met at one point, the answer to the problem would be complete since the resultant must pass through the same point. In this case, there is no common meeting point of the forces and the position of the resultant on the body must be determined. This is done by placing a point O (the polar point) *anywhere* inside or outside the force polygon and connecting it to the letters *a*, *b*, *c*, etc., by polar lines.

A line parallel to the polar line *ao* is now drawn on the free-body diagram *anywhere across space* A to cut force AB. Then a line parallel to *bo* is drawn *across space* B to cut force BC and so on. Where the last line (parallel to the polar line *fo*) intersects the first line (parallel to polar line *ao*) is one point on the line of action of the resultant (or equilibrant). The resultant can now be shown in its position on the free-body diagram by drawing a line parallel to the closing line *af* of the force polygon. In this construction the link lines form a polygon on the forces of the free-body diagram and it is known as a **link polygon.**

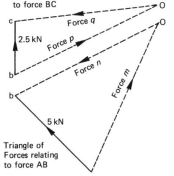

Triangle of forces relating
to force BC

Triangle of
Forces relating
to force AB

Figure 3.10

The graphical conditions of equilibrium for a system of non-concurrent forces are

a) the force polygon must be a closed one;
b) the link polygon must be a closed one.

The validity of the construction of the link polygon can be proved as follows.

Consider the two forces AB and BC of 5 kN and 2.5 kN respectively, which are shown to a larger scale in Fig. 3.10 than in Fig. 3.9.

Any two forces *m* and *n* can be substituted for the one force AB without altering the state of equilibrium of the body. The value of these two components *m* and *n* which replace force AB can be obtained from a triangle of forces (Fig. 3.10). The force of 5 kN represented by the line *ab* is the resultant of the two forces. In a similar manner, force BC can be replaced by two forces *p* and *q*; the values of these two forces also being given by a triangle of forces.

If it is arranged that component *n* of force AB is equal to component *p* of force BC, the net result on the body of these two components is nil, since they act in the same straight line but in the opposite direction (see Fig. 3.10). Similarly, by having a common polar point in the force diagram (given again in Fig. 3.11) component *q* of force BC will balance with an equal component of force CD and so on. It follows that, assuming the equilibrant acts through the point where the two link lines parallel to polar lines *ao* and *fo* meet, the net result on the body of all the forces, including the equilibrant, is nil. Therefore the construction is valid.

Note: the arrow representing the direction of the equilibrant will be in the opposite direction to that given for the resultant in Fig. 3.9.

In brief, this method of dealing with non-concurrent forces consists in replacing each of the given forces by two forces or components and arranging that one component of each force is equal in value to a component of the succeeding force. This can be done by having a common polar point in the force diagram. If the polar point is at any other position different to that shown in Figs. 3.9 and 3.11, the construction is still valid, since a force can be replaced by two components in a multitude of ways. Although the polar point can theoretically be placed anywhere, a neater solution can be obtained by choosing its position so that the link polygon will fit nicely on the free body diagram. A multitude of link polygons can be drawn for any one problem, by using different positions of the polar point, but the closing lines (such as lines drawn parallel to *ao* and *fo* in Fig. 3.9) will all intersect on points along the line of action of the equilibrant (or resultant).

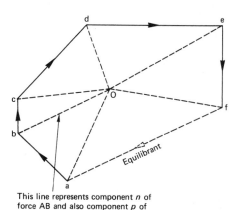

This line represents component *n* of
force AB and also component *p* of
force BC

Figure 3.11

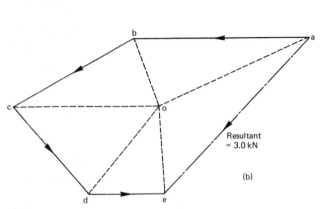

Figure 3.12 (*a*) Free-body diagram with link polygon; (*b*) Force polygon with polar diagram

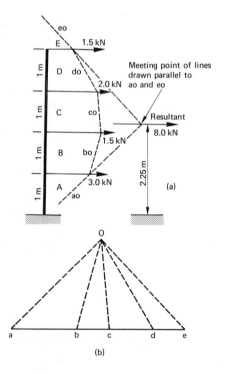

Figure 3.13 (*a*) Free-body diagram with link polygon; (*b*) Force and polar diagrams

APPLICATION OF THE LINK POLYGON

Example 3.4

A vertical post is subjected to pulls from four cables as shown in Fig. 3.12. The magnitude, direction and position of the resultant pull is obtained as shown.

Note that in Fig. 3.12(*a*) line *ao* has to be produced beyond space A until it intersects line *eo*.

Example 3.5

A vertical post it subjected to parallel pulls from four cables as shown in Fig. 3.13. Determine the position of the resultant pull.

The resultant pull is 8 kN acting at a point 2.25 m from the foot of the post as shown in Fig. 3.13.

It is usually quicker and easier to obtain the resultant of parallel forces by calculation (see Chapter 4).

The link polygon may also be used to determine the reactions to a beam or truss. It must be realised, however, that the values of two unknown forces in a system of non-concurrent forces which are in equilibrium can only be determined when the following conditions are satisfied:

1 All the forces (apart from the two unknowns) must be known completely, that is, in magnitude, line of action and direction.

2 The line of action of one of the unknown reactions (not necessarily its direction) must be known.

3 At least one point on the line of action of the remaining reaction must be known.

Example 3.6

A horizontal beam is hinged at one end and supported on a smooth wall at the other end. It is loaded as indicated in Fig. 3.14 (the central load including the weight of the beam). Determine the reactions at the supports.

Figure 3.14

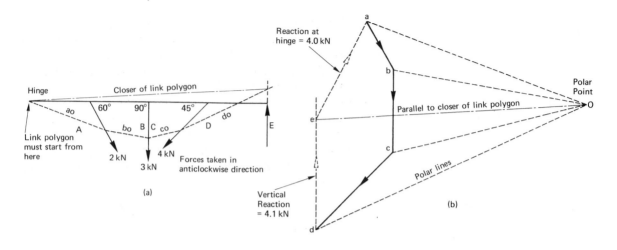

Figure 3.15 (a) Free-body diagram with link polygon; (b) Force and polar diagrams

Solution Since the wall is smooth, the reaction at the wall is at right angles to the surface and is therefore vertical. The line of action of one of the reactions is thus known and so is one point (the hinge) on the line of action of the remaining reaction. It is not necessary, for the solution, to know beforehand the direction of the arrow on the vertical reaction, but is should be obvious that in this example the reaction is acting in an upward direction.

The free-body diagram is drawn to scale and the three known forces are plotted on the force diagram, i.e. *a* to *b*, *b* to *c*, and *c* to *d*, as shown in Fig. 3.15. From *d* a vertical line is drawn parallel to the force DE (the unknown vertical reaction). The point *e* will be somewhere on this line, and its position cannot be fixed until the link polygon is drawn.

Choose a polar point O and draw the polar lines. *From the hinge* draw a line across the A space parallel to polar line *ao*. Where this line cuts the force AB, draw a line across the B space parallel to polar line *bo* and so on, the last line being drawn parallel to the polar line *do* to cut the vertical reaction (force DE). From this point on the force DE draw a line to the hinge. This line closes the link polygon and is known as the *closer*. Draw a line on the force diagram from the polar point O and parallel to the closer. Where this line cuts the vertical line from *d* is the position of *e*. Join *e* to *a*. The length of line *de* gives the magnitude of the vertical reaction at the right-hand support (4.1 kN) and the length of line *ea* gives the magnitude and direction of the reaction at the hinge (4.0 kN).

It is absolutely essential that the link polygon is started by drawing a line from the *only known point* on the reaction at the hinge. Referring to Fig. 3.10 and the procedure necessary to obtain the equilibrant of a system of non-concurrent forces, it was shown that lines drawn parallel to the first and last polar lines intersect on one point of the line of action of the equilibrant. In the procedure just described for obtaining two unknown forces, advantage has been taken of the fact that one point on the unknown reaction is known, and therefore it can be arranged that lines drawn parallel to the first and last polar lines meet at this point (the last polar line being the closer).

Figure 3.16

Example 3.7

A beam as shown in Fig. 3.16 is supported by two walls and loaded as shown, the central load including an allowance for the weight of the beam. Determine the value of the reactions.

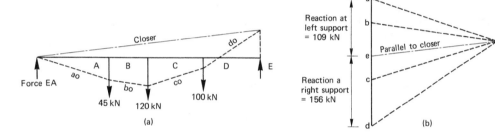

Figure 3.17 (*a*) Free-body diagram with link polygon; (*b*) Force and polar diagrams

Solution Solving a problem of this type (where all the forces are vertical) by means of the link polygon is equivalent to using a sledge hammer to drive in a tintack. Solution by calculation is much quicker and easier (Chapter 5). The graphical solution is shown in Fig. 3.17. The reaction at the right support is 156 kN (given by *de* on the force diagram) and the reaction at the left support is 109 kN (given by *ea*).

Example 3.8

Fig. 3.18 represents a roof truss one end of which is supported by a roller bearing. Determine the reactions. The vertical forces are due to the weight of the roof covering, etc., and the inclined forces (which are at right angles to the slope of the roof) are due to wind loads.

Figure 3.18

Free body diagram

(a)

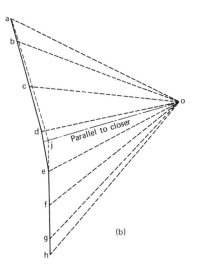

(b)

Figure 3.19 (a) Free-body diagrams with link polygon; (b) Force and polar diagrams

Solution The manner in which the reactions are determined is shown in Fig. 3.19. It should be noted that the right-hand reaction is vertical because the rollers, as indicated, cannot resist horizontal forces, and although both reactions can be determined without combining the pairs of forces by the parallelogram law, the solution given here is probably the most straightforward.

Procedure. The free-body diagram is drawn and the parallelogram of forces law used to combine the vertical and inclined forces. The known loads AB, BC, etc., are plotted in the force diagram, the last load to be plotted being force GH (represented by *gh* on the force diagram). Since the reaction at the right support (force HJ) is vertical, the point *j* will be somewhere on the vertical line drawn from *h* and the link polygon will be the means of fixing this point *j*.

It will be observed that two polar lines (*ao* and *ho*) appear not to be used for constructing the link polygon, and this fact needs explanation. The link polygon must start from the hinge which is the only known point on the line of action of the reaction JA. Going clockwise, the next force to the reaction is the force AB. Therefore from the hinge a line must be drawn parallel to *ao* across the A space to cut force AB. But force AB passes through the hinge. Therefore a line parallel to *ao* must be drawn *from* the hinge *to* the hinge. A line drawn parallel to polar line *ao* may therefore be considered as contained in the hinge.

The next force in order is BC, so a line is drawn from the hinge (which is a point on force AB as well as a point on the reaction) parallel to polar line *bo* to cut force BC and so on, finishing with a line drawn parallel to polar line *go* to cut force GH.

The next force to GH is the vertical reaction HJ so a line parallel to polar line *ho* must be drawn from force GH to cut force HJ. But these two forces are in the same straight line, so, in effect, the line drawn parallel to polar line *ho* from force HG to force HJ is of zero length.

When the point *j* is fixed by drawing a line parallel to the closer of the link polygon, the line *hi* gives the reaction at the right support (34 kN approximately) and the line *ia* gives the direction and magnitude (38 kN approximately) of the reaction at the left support.

Many roof trusses, particularly those of small spans, are equally supported or hinged at both ends so that neither of the reactions is actually vertical. Even so, for the purposes of determining the reactions and the forces in the members of the truss, it is frequently assumed that one reaction is vertical. An alternative assumption with respect to the reactions is that, if the ends of the truss are similarly supported, the horizontal components of the two reactions are equal. This implies that each support resists equally the tendency to horizontal movement due to any inclined loads on the truss.

Example 3.9

A truss as shown in Fig. 3.20 is subjected to wind forces and dead loads. Determine the magnitudes and directions of the reactions at the supports assuming that the total horizontal component of the applied loads is shared equally by the two supports.

Procedure. After combining by the parallelogram of forces, the loads are plotted, finishing with *de* on the force diagram. The perpendicular line *ax* gives the sum of the vertical components of all the loads on the truss, and the horizontal line *xe* gives the sum of the horizontal components of the loads. The total horizontal load on the truss is therefore given by *xe* and, since this is resisted equally by the two supports, the horizontal component of each reaction is given by half the length of line *xe*. Point *f* can now be fixed, *ef* representing the horizontal component EF of the reaction at the right support.

The next force in order is the vertical component FG, so from *f* a vertical line is drawn. The position of *g* on this line is not known but it is known that the vertical component FG plus the vertical component GH must equal the total

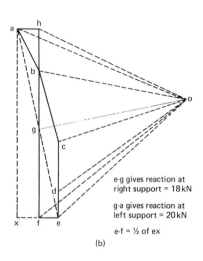

e-g gives reaction at
right support = 18 kN

g-a gives reaction at
left support = 20 kN

e-f = ½ of ex

(b)

Figure 3.20 (*a*) Free-body diagram with link polygon; (*b*) Force and polar diagrams

vertical component of the applied loads. Point *h* must therefore be in a horizontal line from *a* and the force polygon is completed by the line *ha* which represents the horizontal component HA of the reaction at the left support.

The problem is now reduced to determining the vertical components of the two reactions—in other words, to fixing the position of point *g* on the vertical line *fh*.

Starting the link polygon at the left hinge, the first effective link line is the one drawn parallel to polar line *bo* across the B space to cut force BC. (The first imaginary link line is *from* the hinge on the unknown vertical component GH to cut the horizontal component HA *at* the hinge, and the second link line is *from* the hinge on force HA to cut the force AB *at* the hinge.) Link lines parallel to *co* and *do* are now drawn.

The next force in order is EF which is horizontal. Therefore the link line parallel to *eo* is drawn across the E space to cut the horizontal force EF. The last link line is drawn parallel to *fo* across the F space to cut the unknown vertical component FG.

Finally, the closer connecting the two unknown vertical components of the reactions is drawn, and a line parallel to the closer is drawn on the force diagram from O. Where this line cuts the vertical line *fh* is the position of *g*. Then *fg* gives to scale the vertical component of the reaction at the right support and *gh* gives the vertical component of the reaction at the left support. The actual magnitudes and directions of the reactions are given by lines *eg* and *ga* respectively.

Figure 3.21

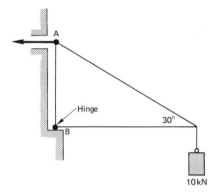

Figure 3.22

EXERCISE 3

1 Neglecting the weight of the pole, determine the reactions at A and B, Fig. 3.21.

2 Neglecting the weights of the members, determine the reactions at A and B of Fig. 3.22. (Point A is supported by a horizontal cable.) In addition, determine the forces in the three members of the framework, and the horizontal and vertical components of the reaction at B.

3 A gate weighing 500 N is supported by two hinges so that the lower hinge takes the whole weight of the gate (Fig. 3.23). Find the reactions at the hinges, and the vertical and horizontal components of the reaction at the bottom hinge.

4 Determine the tension in the tie cable and the reaction at the hinge (Fig. 3.24).

5 Determine the tension in the cable and the reaction at the hinge (Fig. 3.25).

6 A ladder rests against a smooth sloping wall, as shown in Fig. 3.26. The vertical load includes the weight of the ladder. Determine the reactions at A and B.

7 The structure shown in Fig. 3.27 is supported by a smooth ring bolt at A and by a socket at B. Determine the reactions at A and B. Determine also the horizontal and vertical components of the reaction at B.

Figure 3.23

Figure 3.24

Figure 3.25

Figure 3.26

Figure 3.27

Figure 3.28

Figure 3.29

Figure 3.30

8 Determine the tension in the string and the reaction at the hinge (Fig. 3.28). Neglect the weight of the members.

9 Determine the tension in the tie rope and the reaction at the hinge (Fig. 3.29).

10 A uniform beam weighing 500 N is supported horizontally as shown in Fig. 3.30. Determine the reactions at the smooth wall and hinge.

11 The horizontal cable of Fig. 3.31 is half-way up the rod, the weight of which may be neglected. Determine the tension in the cable and the reaction at the hinge.

12 A uniform rod is hinged half-way along its length as shown in Fig. 3.32. Determine the tension in the tie rope and the reaction at the hinge, neglecting the weight of the rod.

13 A uniform rod is hinged half-way along its length as shown in Fig. 3.33. Determine the tension in the rope and the reaction at the hinge, neglecting the weight of the rod.

14 Determine the resultant of the two forces of 2.5 kN and 1.5 kN (Fig. 3.34), then determine the tension in the cable and the reaction at the bottom of the pole. Neglect the weight of the pole.

Figure 3.32

Figure 3.31

Figure 3.33

Figure 3.34

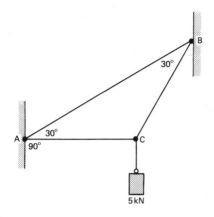

Figure 3.35

15 A framework consisting of three rods hinged at the joints is supported by a smooth wall at B and is hinged to a wall at A (Fig. 3.35). Determine the reactions at A and B and the forces in the three members, neglecting their weight.

16 A bent rigid rod has two equal arms AC and BC as shown in Fig. 3.36. The wall at A is smooth. Determine the angle x so that the reaction at B is at right angles to the surface of the wall.

17 Determine the tensions in the strings and the reaction at the hook (Fig. 3.37).

18 A horizontal beam, the weight of which may be neglected, is loaded half-way along its length as shown in Fig. 3.38. Determine the reactions at A and B.

19 Neglecting the weight of the beam, determine the reactions at A and B of Fig. 3.39.

20 Neglecting the weight of the rod, determine the reactions at A and B of Fig. 3.40.

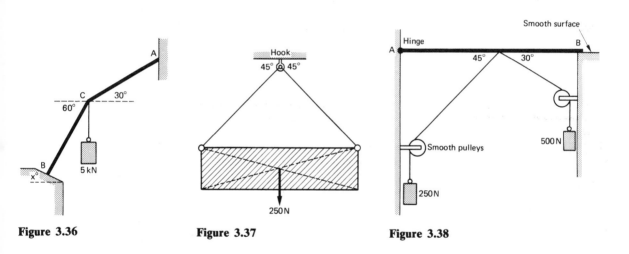

Figure 3.36   **Figure 3.37**   **Figure 3.38**

Figure 3.39   **Figure 3.40**

Figure 3.41

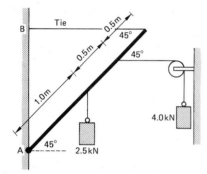

Figure 3.42

21 A horizontal rod the weight of which may be neglected is hinged at X as shown in Fig. 3.41. Determine by the parallelogram of forces the resultant pull on the smooth pulley, then use the triangle of forces to determine the tension in the tie rope and the reaction at the hinge.

22 A rod, the weight of which may be neglected, is hinged at A and supported by a tie rope connected to B as in Fig. 3.42. Determine the resultant pull due to the two weights of 2.5 kN and 4 kN, then determine the forces in the tie and rod.

23 Determine the reactions A and B to the rod shown in Fig. 3.43.

24 Determine the position of the resultant for the systems of forces given in Fig. 3.44(a), (b) and (c).

25 A mast, fixed at its base is loaded as shown in Fig. 3.45. Determine the magnitude and line of action of the resultant of the given loads.

26 Determine the magnitude and line of action of the resultant of the four parallel forces acting on the mast of Fig. 3.46.

Figure 3.43

Figure 3.45   **Figure 3.46**

Figure 3.44

Figure 3.47

Figure 3.48

Figure 3.49

Figure 3.50

27 Determine the magnitude of the vertical reaction at R and the magnitude and line of action of the reaction at L of the system of forces given in Fig. 3.47.

28 A beam (Fig. 3.48) is hinged at L and supported by means of a cable at R. Determine the reactions at L and R.

29 A vertical mast is supported by means of a smooth ring bolt at X (which means that the reaction at X is horizontal) and by a socket at Y (Fig. 3.49). Determine the reactions at X and Y.

30 A tower (Fig. 3.50) is supported by roller bearings at R and is hinged at L. Determine the reactions at L and R.

31 Determine the reactions at L and R due to the three vertical loads of Fig. 3.51.

32 Assuming the reaction at R of Fig. 3.52 to be vertical, determine its magnitude. Also determine the magnitude and direction of the reaction at L.

33 Repeat Question 32 assuming that the total horizontal component of the applied loads is shared equally by the two supports.

34 The reaction at R to the truss shown in Fig. 3.53 is assumed to be vertical. Determine its value due to the wind loading given and also determine the value and line of action of the reaction at L.

35 Repeat Question 34 assuming that the total horizontal component of the applied loads is shared equally by the two supports.

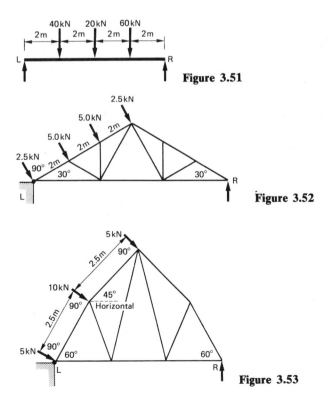

Figure 3.51

Figure 3.52

Figure 3.53

36 Determine the reactions at L and R of Fig. 3.54 assuming the reaction at R to be vertical.

37 Repeat Question 36 assuming that the total horizontal component of the applied loads is shared equally by the two supports.

38 The lengths of rafters represented by x and y of Fig. 3.55 are equal and the wind loads are at right angles to the slope of the roof. Determine the reactions at L and R assuming the reaction at L to be vertical.

39 Fig. 3.56 shows a truss which is supported at L and R. Determine the reactions, assuming that the total horizontal component of the applied loads is shared equally by the two supports.

Figure 3.54

Figure 3.55

Figure 3.56

4 Moments of Forces

The concept of the turning effect or moment of forces is explained in this chapter. The principle of the moment is then applied to the solution of problems connected with the equilibrium of a system of non-concurrent forces, including parallel forces (e.g. beam reactions, couples, etc.).

It is common knowledge that a small force can have a big turning effect or leverage. In mechanics, *moment* is commonly used instead of *turning effect* or *rotational effect*. The word moment is usually thought of as having some connection with time. Moment, however, derives from a Latin word meaning movement and moments of time refer, of course, to movement of time. Similarly, the **moment of a force** can be thought of as movement of a force, although a more exact definition is "turning effect of a force". In many types of problem, there is an obvious **turning point** (or hinge or pivot or fulcrum) about which the body turns or tends to turn as a result of the effect of the force (see Fig. 4.1 (*a*) to (*e*)). In other cases, where the principle of moments may be used to obtain solutions to problems, there may not be any obvious turning points.

To solve problems by the use of the principle of moments, the effect of the forces must be considered in relation to the direction in which the body turns or tends to turn around the given point. In Fig. 4.1(*a*) the turning effect or moment of the force is anticlockwise; in (*b*) the moment is clockwise; and in (*c*) and (*d*) the moments are anticlockwise. There are two turning points in Fig. 4.1(*e*). Considering the point where the crowbar touches the ground, (point (1)), the moment of force at A is clockwise. At B, the bar applies a force to the stone and causes it to turn about point (2). The moment of force applied to the stone at B is therefore anticlockwise.

If the line of action of a force passes through a point it can have no turning effect (moment) about that point. For example, a door cannot be opened by pushing at the hinge.

Further examples are given in Fig. 4.2. The given forces have no moments about points O.

MEASUREMENT OF MOMENTS

Fig. 4.3 represents a uniform rod which balances on a fulcrum placed half-way along the length of the rod. If a force of 25 N is suspended at 0.3 m from the fulcrum it will be found that a force of 15 N is required at 0.5 m from the fulcrum on the other side in order to maintain equilibrium. Instead of 15 N at 0.5 m from the fulcrum we could place 12.5 N at 0.6 m or 7.5 N at 1.0 m. The results of the experiments would show that the force causing a clockwise moment multiplied by its distance from the turning point equals the force causing an anticlockwise moment multiplied by its distance from the turning point. For example, referring to Fig. 4.3, the clockwise moment of the 15 N force is 15 N × 0.5 m or 7.5 N m.

Moments are always expressed in force-length units such as N m, N mm, kN m, etc. A little consideration should show that this must be so, because the turning effect of a force depends on distance as well as on the magnitude of the force.

The distance from the force to the turning point is often called the **lever arm** of the force. In Fig. 4.3 the lever arm of the 15 N force is 0.5 m and the lever arm of the 25 N force is 0.3 m. It should be noted that the lever arm should be measured from the turning point to the point where it intersects the line of action of the force at right angles. In Fig. 4.3 the lines of action of the weights are vertical and the lever arms are horizontal. In Fig. 4.4 the lever arms of the forces are AO and BO respectively and for equilibrium $T \times AO$ must equal $P \times BO$. The reason for measuring the lever arms in this way will be made clear in Example 4.3 on page 45.

Further demonstrations of the principle of moments can be arranged quite easily with the aid of a fulcrum, a rod and a few weights. In Fig. 4.5 the rod itself, since it is supported at its centre of gravity, has no resultant turning effect, clockwise or anticlockwise.

$$\text{Anticlockwise moments} = \text{Clockwise moments}$$
$$(75 \times 300) + (45 \times 240) = (150 \times 150) + (30 \times 360)$$
$$33\,330 \text{ N mm} = 33\,300 \text{ N mm}$$

Example 4.1

A plank of uniform cross-section 4 m long and weighing

Figure 4.1

Figure 4.2

Figure 4.3

Figure 4.4

Figure 4.5

1.2m   2.8m
0.8m
c.g.

300N

W
200

Figure 4.6

30   30

30   30   30   30   30   30

Loads are in newtons

Figure 4.7

300 N has a support placed under it at 1.3 m from one end as shown in Fig. 4.6. Calculate the magnitude of the weight W required to cause the plank to balance.

Solution It was stated on page 23 that, when considering the equilibrium of forces, the whole weight of a body can be *assumed* to act at its centre of gravity. In the case of a uniform rod or beam, this is half-way along the rod. Taking moments about the fulcrum or turning point,

$$1.2\,W\,N\,m = 300 \times 0.8\,N\,m$$

$$W = \frac{240\,N\,m}{1.2\,m} = 200\,N$$

The whole weight of the plank is not, of course, concentrated at its centre of gravity. Each little particle of the plank is being attracted downwards by the earth and the 1.2 m length of the plank to the left of the fulcrum is exerting an anticlockwise moment, whilst the 2.8 m length of the plank to the right of the rod is exerting a clockwise moment. The total net effect is, however, the same as if the whole of the weight were acting at 0.8 m to the right of the fulcrum.

For example, assume the plank is divided into 0.4 m lengths each length weighing 30 N, and assume that each 30 N weight acts half-way along its 0.4 m length (Fig. 4.7).

Taking moments about the fulcrum

$$30 \times (0.2 + 0.6 + 1.0) + 1.2\,W$$
$$= 30 \times (0.2 + 0.6 + 1.0 + 1.4 + 1.8 + 2.2 + 2.6)$$

$$(54 + 1.2\,W)\,N\,m = 294\,N\,m$$
$$1.2\,W\,N\,m = 240\,N\,m$$
$$W = 200\,N$$

If the rod is divided into a thousand lengths each length weighing 0.3 N, and moments are taken about the fulcrum, the answer will be the same, so there need be no hesitation in assuming the whole weight of a body to act at its centre of gravity.

Example 4.2

A uniform rod weighing 50 N and carrying weights as shown (Fig. 4.8) is hinged at A. The rod is supported in a horizontal

String

200 mm   175 mm   225 mm

A

B

60 N   20 N

Figure 4.8

Figure 4.9

position by a vertical string at B. Calculate the tension in the string and the reaction at the hinge.

Solution The free-body diagram is given in Fig. 4.9.

If the string were to be cut, all the weights would pull the rod downwards in a clockwise direction about the hinge. The string exerts a counterbalancing anticlockwise moment.

Taking moments about the hinge, assuming T newtons to be the tension in the string,

$$375T = (60 \times 200) + (50 \times 300) + (20 \times 600)$$
$$375T = 39\,000$$
$$T = 39\,000/375 = 104\,\text{N}$$

The total downward force is $60 + 50 + 20$, i.e. 130 N. Since the string is only holding up with a vertical force of 104 N, the reaction at the hinge must be $130 - 104$, i.e. 26 N acting upwards.

Example 4.3

A horizontal rod has a weight of 100 N suspended from it as shown in Fig. 4.10. Calculate the tension in the string and the reaction at the hinge, ignoring the weight of the rod.

Figure 4.10

Solution 1 Let the tension in the string be represented by T newtons and resolve this force into its vertical and horizontal components. This may be done by calculation or by drawing a line 10 units long to represent T (Fig. 4.11).

It is found that the vertical component V is 5 units long, i.e. $0.5T$, and that the horizontal component H is 8.66 units long, i.e. $0.866T$. The line of action of the horizontal component of the tension in the string passes through the hinge and therefore has no moment about this point. Taking moments about the hinge

Figure 4.11

Anticlockwise moment = Clockwise moment
$$0.5T \times 1\,\text{m} = 100 \times 0.5\,\text{m}$$
$$0.5T = 50$$
$$T = 50/0.5 = 100\,\text{N} = \text{tension in string}$$

Reaction at hinge: the total downward force is 100 N. The vertical component of the tension in the string is $0.5T$, i.e. 50 N. (The horizontal component of the tension in the string is ineffective in holding up the suspended weight.) Since there is 100 N pulling down, there must be 100 N pushing up, therefore the hinge must supply a vertical reaction in the upward direction of 50 N. But the string has also a horizontal component of $0.866T$, i.e. 86.6 N, tending to pull the rod away horizontally from the hinge. The hinge must resist this

Figure 4.12

Figure 4.13

Figure 4.14

pull. Therefore there must be a horizontal force acting at the hinge in addition to a vertical force.

The total reaction at the hinge is given in Fig. 4.12 and may be found from a scale drawing, or, if R is the reaction at hinge,

$$R^2 = 50^2 + 86.6^2 = 10\,000$$
$$R = \sqrt{10\,000} = 100\ N$$

Solution 2 In Fig. 4.13 which may be drawn to scale, produce the line of action of the string and draw a line AO at right angles, then taking moments about O,

$$T \times OA = 100 \times 0.5$$
$$T \times 0.5 = 50$$
$$T = 100\ N$$

This method is not so convenient as Solution 1 for determining the reaction at the hinge.

Solution 3 Since there are only three forces in this problem, the triangle of forces law can be applied (see page 9), and this method may be quicker for problems in which the forces are inclined.

CONDITIONS OF EQUILIBRIUM

Example 4.4

A ladder rests against a smooth vertical wall and a man weighing 750 N stands on it as indicated in Fig. 4.14. Neglecting the weight of the ladder, determine the reactions at the wall and at the ground.

Solution This problem is identical with that on page 24 but it is now being solved by using the principle of moments. Since the wall is smooth, the reaction at A is horizontal. There may not appear to be any obvious turning point but B can be considered to be a turning point. (If the wall were to collapse the ladder would swing round B in an anticlockwise direction due to the force of 750 N. If one could imagine the force at A to be increased above that necessary to maintain equilibrium, it would turn the ladder round B in a clockwise direction.) By taking moments about B, the reaction at B can be ignored since it passes through this point and therefore has no moment about it.

Reaction at $A \times$ distance $CB = 750\,N \times$ distance DB

$$R_A \times 6 = 750 \times 1$$

$$R_A = 750/6 = 125\,N$$

Reaction at B (R_B): since the wall is smooth, it cannot help at all in holding *up* the ladder, therefore the whole weight of 750 N is taken at B. There is therefore an upward reaction at B of 750 N. There must also be a horizontal reaction to prevent the foot of the ladder from moving outwards. This horizontal reaction equals the force at A, i.e. 125 N, therefore the total reaction at B which can be obtained graphically or by calculation is 760 N.

The above example is an illustration of the application of the three **laws of equilibrium** which apply to a system of forces acting in one plane. These laws are

1 The algebraic sum of the vertical forces must equal zero, i.e. if upward forces are called positive and downward forces are called negative, then a force of +750 N must be balanced by a force of −750 N and the algebraic sum of the two forces is equal to zero. This can also be expressed by

$$\sum V = 0$$

where $\sum$ is the Greek letter "sigma" (letter S) and in this connection means "the sum of", whilst V represents "the vertical forces", upward forces being plus and downward forces being minus.

2 The algebraic sum of the horizontal forces must equal zero, i.e. the sum of the horizontal forces (plus forces) acting towards the left must equal the sum of the horizontal forces (minus forces) acting towards the right or

$$\sum H = 0$$

3 The algebraic sum of the moments of the forces must equal zero, i.e. if clockwise moments are called plus and anticlockwise moments are called minus then

$$\sum M = 0$$

Example 4.5

A vertical pole (Fig. 4.15) is supported by being hinged in a socket at B and by a smooth ring which can give no vertical reaction, at A. Calculate the reactions at A and B neglecting the weight of the pole.

Figure 4.15

Figure 4.16

Figure 4.17

Figure 4.18

Solution First, resolve the pull of 5 kN into its vertical and horizontal components as shown in Fig. 4.16 and take moments about B.

$\sum M = 0$ or clockwise moments equal anticlockwise moments.

$$4.33 \times 4.5 = R_A \times 1.5$$
$$R_A = 13.0 \text{ kN}$$

(Note that the vertical component of 2.5 kN has no moment about B since its line of action passes through B.)

$$\sum V = 0$$

The vertical component of the reaction at B must equal the vertical component of the pull at the top of the pole since the support of A can only *hold back* and cannot *hold up*. The vertical component of the reaction at B therefore equals 2.5 kN and acts upwards.

$$\sum H = 0$$

The horizontal force at the top of the mast is 4.33 kN acting → and the reaction at A is acting ←. There must be, therefore, a horizontal force at B acting → equal to $13.0 - 4.33$, i.e. 8.67 kN. The complete reaction at B is given by Fig. 4.17. Note that, if the pole itself weighs, say, 0.5 kN, this weight would be taken entirely at B (since the pole is vertical) and the vertical component of the reaction at B would be 3 kN instead of 2.5 kN as shown in Fig. 4.17. The complete reaction at B would then be

$$R_B = \sqrt{[8.67^2 + 3^2]} = 9.2 \text{ kN}$$

Another method of finding the reaction at A (neglecting the weight of the pole) is shown in Fig. 4.18.

$$5 \text{ kN} \times \text{distance } BC = R_A \times 1.5 \text{ m}$$

Note also that the triangle of forces principle could be applied to determine both reactions.

Example 4.6

A cantilever truss (Fig. 4.19) is supported at A and B in such a manner that the reaction at A is horizontal. Neglecting the weight of the truss, calculate the reactions at A and B.

Solution Take moments about B to determine the reaction at A.

$$\sum M = 0$$
$$1.5R_A = 10 \times 1.5 + 5 \times 3$$
$$R_A = 20 \text{ kN acting towards the right.}$$

Figure 4.19

Figure 4.20

Figure 4.21

Figure 4.22

$\sum H = 0$, therefore horizontal component of reaction at B is 20 kN acting towards the left.

$\sum V = 0$, therefore vertical component of reaction at B is 15 kN acting upwards.

Referring to Fig. 4.20

$$R_B = \sqrt{[20^2 + 15^2]} = \sqrt{625} = 25 \text{ kN}$$

RESULTANT OF PARALLEL FORCES

It was stated in Chapter 2 in connection with concurrent forces that the resultant of a given number of forces is the single force which, when substituted for the given forces, has the same effect on the *state of equilibrium* of the body. This definition applies also to parallel forces.

Consider Fig. 4.21 which represents a beam of negligible weight supporting three vertical loads of 13, 18 and 27 kN respectively.

The force in the vertical cable can be obtained by taking moments about the hinge, i.e.

$$F \times 3 \text{ m} = 13 \times 1.4 + 18 \times 2.4 + 27 \times 3.3$$
$$= 150.5 \text{ kN m}$$
$$F = 150.5/3 = 50.2 \text{ kN}$$

Now, if one force (the resultant) is substituted for the three downward loads without altering the state of equilibrium, its magnitude must be $13 + 18 + 27 = 58$ kN, and if the reaction in the cable is to remain as before, i.e. 50 kN, the moment of the single resultant force about the hinge must equal the sum of the moments of the forces taken separately.

Referring to Fig. 4.22, moment of the resultant force about the hinge is $58x$ kN m, and this must equal 150.5 kN m, therefore

$$58x = 150.5$$
$$x = 150.5/58 = 2.59 \text{ m}$$

The resultant of the three forces is, therefore, a single force of 58 kN acting at 2.59 m from the hinge. The reactions in the cable and at the hinge will be the same for this one force as for the three forces.

Example 4.7

Calculate the magnitude and position of the resultant of the system of forces shown in Fig. 4.23 and determine where a

Figure 4.23

Figure 4.24

Figure 4.25

fulcrum must be placed to maintain equilibrium. (Neglect the weight of the rod.)

Solution This in an example of a system of *unlike parallel forces* since all the forces are not acting in the same sense. The moment of this resultant about *any point* must equal the algebraic sum of the moments of the separate forces. Taking moments about end A,

$$150x = 25 \times 2 - 50 \times 3 + 75 \times 4$$
$$= 50 - 150 + 300$$
$$x = 200/150 = 1.33 \text{ m}$$

The fulcrum must therefore be placed at this point in order to maintain the rod in balance (and the reaction at the fulcrum $= 150$ N).

The same answer would be obtained by taking moments about any point, as for example, the point B, 2 m from A (see Fig. 4.24).

$$150x = 100 \times 2 + 25 \times 4 - 50 \times 5 + 75 \times 6$$
$$= 200 + 100 - 250 + 450$$
$$x = 500/150 = 3.33 \text{ m}$$

The position of the resultant is therefore 3.33 m from B or 1.33 m from A.

Problems can often be simplified by working with a resultant instead of with a number of separate forces. This is demonstrated in the following example.

Example 4.8

Calculate the resultant of the three downward forces of Fig. 4.25, then use the triangle of forces principle to determine the reaction in the cable and the reaction at the hinge.

Solution The resultant of the three downward forces is 6 kN and, taking moments about the hinge,

$$6x = 3 \times 0.6 + 1 \times 1.2 + 2 \times 2.5$$
$$x = 8/6 = 1.33 \text{ m}$$

Now, using the method of Chapter 3, page 24, the reactions can be found as shown in Fig. 4.26.

COUPLES

Two equal unlike parallel forces are said to form a **couple.**
Imagine a rod resting on a smooth table, as shown in Fig.

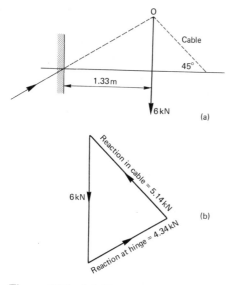

Figure 4.26 (*a*) Free-body diagram; (*b*) Force diagram for point O

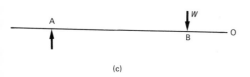

Figure 4.27

4.27(*a*), and assume that two equal unlike parallel forces of 20 N are applied horizontally as shown. The rod will rotate in a clockwise direction, and the moment causing this rotation is obtained by multiplying *one* of the forces by the distance AB between the two forces. The perpendicular distance AB is called the *arm of the couple* or in general terms, the **lever arm,** l_a. Therefore

$$\text{Moment of the couple} = 20\,\text{N} \times 150\,\text{mm}$$
$$= 3000\,\text{N mm}$$

No matter about what point the moment is considered, the answer is equal to $W \times AB$. For example, taking moments about any point O between the two forces (Fig. 4.27(*b*)), the moment causing rotation is

$$(W \times AO) + (W \times BO)$$

This equals

$$W(AO + BO) = W \times AB$$

In Fig. 4.27(*c*) the clockwise moment about O is $W \times AO$, and the anticlockwise moment about O is $W \times BO$.

The net moment causing rotation is

$$W \times AO - E \times BO = W(AO - BO)$$
$$= W \times AB$$

Note that a couple acting on a body produces rotation and the forces cannot be balanced by a single force. To produce equilibrium another couple of equal and opposite moment is required.

Fig. 4.28 represents a mast fixed at its base. The moment tending to break the mast at its base is

$$2(x + 1) - 2x \text{ kN m} = 2x + 2 - 2x = 2 \text{ kN m}$$

No matter what the distance x may be, the moment at the base will be 2 kN m. This moment or couple is resisted by the fixing moment or couple applied by the base to the mast.

A knowledge of the principles of couples is useful in understanding beam problems (see Chapter 9).

BEAM REACTIONS

The application of the principle of moments of forces to the calculation of beam reactions can be verified by means of the following experiment.

A beam of wood, a little longer than 1.2 m and weighing 15 N, has weights suspended from it as shown in Fig. 4.29. It

Figure 4.28

Figure 4.29

Figure 4.30

Figure 4.31

is required to determine the reactions at supports A and B.

If a spring balance is attached to the beam at B, the beam can be lifted just clear of the support and it will be observed that the reading on the balance is 32.5 N. When the balance is released so that the beam rests again on block B, it follows that the beam must be pushing down on the support with a force of 32.5 N. Similarly, the upward reaction of the support on the beam is 32.5 N.

If the hand holding the balance is moved up and down, the beam will turn round the point where it is supported at A. Taking moments about this point, representing the reaction at B by R_B,

$$R_B \times 1.2 \text{ m (anticlockwise about } A)$$
$$= 35 \times 0.4 + 15 \times 0.6 + 20 \times 0.8$$

therefore

$$1.2 R_B = 39 \quad \text{and} \quad R_B = \frac{39}{1.2} = 32.5 \text{ N}$$

If the balance is transferred to A and the beam lifted off its support, the balance will read 37.5 N (Fig. 4.30).

By calculation, taking B as the turning point,

$$R_A \times 1.2 = 20 \times 0.4 + 15 \times 0.6 + 35 \times 0.8$$

clockwise = anticlockwise

$$R_A = \frac{45}{1.2} = 37.5 \text{ N}$$

Note that the sum of the reactions is 70 N and that this is also the sum of the downward forces ($\Sigma V = 0$).

Note also that to determine the reaction at A, moments are taken about B, i.e. B is considered as the turning point and the moment of each force or load is the force multiplied by its distance from B.

Similarly, to determine the reaction at B, moments are taken about A, i.e. A is considered as the turning point.

The two weights of 35 N and 20 N in the discussion above are **point loads** or **concentrated loads,** since they act at definite points on the beam. The beam itself is a **uniformly distributed load** (udl), since its weight is spread uniformly over its whole length. Fig. 4.31 represents the test beam supporting a block of lead 0.4 m long of uniform cross-section and weighing 30 N. For calculating the reactions at the supports, the whole of this uniformly distributed weight can be assumed to act at its centre of gravity.

Taking moments about A

$$1.2 R_B = 30 \times 0.4 + 15 \times 0.6$$

$$R_B = \frac{21}{1.2} = 17.5 \text{ N}$$

Figure 4.32

Figure 4.33

Similarly, taking moments about B

$$1.2R_A = 15 \times 0.6 + 30 \times 0.8$$

$$R_A = \frac{33}{1.2} = 27.5 \text{ N}$$

$$R_A + R_B = 17.5 + 27.5 = 45 \text{ N} = \text{total load}$$

These calculated values of the reactions can be verified by means of the spring balances.

Instead of drawing the beam every time, a free-body diagram can be drawn as in Fig. 4.32 or Fig. 4.33 where two methods of indicating uniformly distributed loads are shown.

Further experiments can be performed with the simple apparatus shown above by placing the supports at different points. Fig. 4.34 shows the supports placed 0.8 m apart. In this case, to determine experimentally the reaction at A, the spring balance must be connected to a point on the beam directly above A.

Figure 4.34

Taking moments about B, the reaction at A produces a clockwise moment and all the loads anticlockwise moments.

$$0.8R_A = 30 \times 0.2 + 15 \times 0.6 + 20 \times 1.2$$

$$R_A = 39/0.8 = 48.75 \text{ N}$$

Taking moments about A, both the reaction at B and the load of 20 N produce anticlockwise moments.

$$0.8R_B + 20 \times 0.4 = 15 \times 0.2 + 30 \times 0.6$$

$$0.8R_B = 3.0 + 18.0 - 8.0$$

$$R_B = \frac{13}{0.8} = 16.25 \text{ N}$$

$$R_A + R_B = 48.75 + 16.25 = 65 \text{ N}$$
$$= \text{sum of the vertical loads}$$

Practical Cases

In buildings, point loads on beams are usually due to loads from other beams or from columns, and uniformly distributed loads are due to floors, walls and partitions and the weights of the beams themselves.

Figure 4.35

Figure 4.36

In Fig. 4.35, assume that the beam AB weighs 10 kN and that it supports two beams which transmit loads of 60 kN and 90 kN respectively to beam AB. In addition, a column transmits a load of 240 kN. The reactions are usually taken as acting at the middle of the bearing area supplied by the supporting brickwork or column. Fig. 4.36 shows the free-body diagram.

Taking moments about B,

$$6R_A = 90 \times 2.0 + 10 \times 3.0 + 240 \times 3.5 + 60 \times 4.0$$
$$= 180 + 30 + 840 + 240$$
$$R_A = \frac{1290}{6} = 215 \text{ kN}$$

Since the total load on the beam is 400 kN, the reaction at B is $400 - 215$, i.e. 185 kN. It is advisable, however, to determine the reaction at B by taking moments about A and thereby to obtain a check on the calculations.

Example 4.9

A beam weighing 9.6 kN supports brickwork 225 mm thick weighing 20 kN/m³ as indicated in Fig. 4.37. Calculate the reactions at the supports.

Solution Both the brickwork and the beam are uniformly distributed loads.

Figure 4.37

Figure 4.38

$$\text{Volume of brickwork} = 5.0 \times 2.5 \times 0.225 = 2.8125 \text{ m}^3$$
$$\text{Weight of brickwork} = 2.8125 \times 20 \qquad = 56.25 \text{ kN}$$
$$\text{Weight of beam} \qquad\qquad\qquad\quad = \underline{\ 9.60 \text{ kN}}$$
$$\text{Total weight} = 65.85 \text{ kN}$$

The free-body diagram is shown in Fig. 4.38.
  Taking moments about B,

$$3.5 R_A = 65.85 \times 2$$
$$R_A = \frac{131.70}{3.5} = 37.63 \text{ kN}$$
$$R_B = 65.85 - 37.63 = 28.22 \text{ kN}$$

(or R_B can be obtained by taking moments about A).

SUMMARY

The **moment** of a force means the turning effect of a force.
  Moments are considered in relation to a turning point, real or imaginary.
  A moment is always obtained by multiplying a force by a distance and this distance must be measured from the turning point to where it cuts the line of action of the force at right angles.
  Moments are expressed in force-length units such as newton metre (N m), newton millimetre (N mm), etc.
  If the line of action of a force passes through a given point, it can have no moment about that point.
  The **conditions of equilibrium** for a system of forces acting in one plane are:

$$\sum V = 0 \qquad \sum H = 0 \qquad \sum M = 0$$

The **resultant** of a number of parallel forces is the algebraic sum of the forces. Its position can be found by taking moments about *any point* and moment of resultant equals the algebraic sum of moments of the individual forces.
  A **couple** consists of two equal unlike parallel forces. The distance between the forces is called the arm of the couple, and the moment of a couple equals one of the forces multiplied by the arm of the couple.

Figure 4.39

Figure 4.40

Figure 4.41

Figure 4.42

Figure 4.43

Figure 4.44

Figure 4.45

EXERCISE 4

1 A uniform rod is in equilibrium under the action of weights as shown in Fig. 4.39. Calculate the value of W and the reaction at the fulcrum, ignoring the weight of the rod.

2 Calculate the value of x metres so that the uniform rod of Fig. 4.40 will balance.

3 A uniform rod weighing 100 N and supporting 500 N is hinged to a wall and kept horizontal by a vertical rope (Fig. 4.41). Calculate the tension in the rope and the reaction at the hinge.

4 A uniform rod weighing 40 N is maintained in equilibrium as shown in Fig. 4.42. Calculate the distance x and the reaction at the fulcrum.

5 A uniform rod 0.4 m long weighing 10 N supports loads as shown in Fig. 4.43. Calculate the distance x if the rod is in equilibrium.

6 A uniform horizontal beam 1.3 m long and weighing 200 N is hinged to a wall and supported by a vertical prop as shown in Fig. 4.44. Calculate the reactions of the prop and hinge.

7 Calculate the reaction at the prop and hinge in Fig. 4.45. The uniform beam weighs 200 N.

8 A compound lever is shown in Fig. 4.46. Each of the horizontal rods weighs 50 N and the vertical rod weighs 40 N. Calculate the value of the force W which is required for equilibrium, and calculate the reactions at the fulcrums. (First take moments about A to find the force in the vertical rod, then take moments about B.)

9 Solve Questions 1 to 13 inclusive of Exercise 3, Chapter 3, with the aid of the principle of moments.

10 Determine the tension in the chain and the reaction at the hinge (Fig. 4.47). The uniform rod weighs 500 N.

Figure 4.46   **Figure 4.47**

Figure 4.48

Figure 4.49

Figure 4.50

11 The horizontal cable in Fig. 4.48 is attached to the rod half-way along its length. The uniform rod weighs 250 N. Calculate the tension in the cable and the reaction at the hinge.

12 Calculate the tension in the guy rope of Fig. 4.49, then use the triangle of forces law to determine the forces in the mast, tie and jib. Neglect the weight of the members.

13 A uniform rod weighing 250 N is loaded as shown in Fig. 4.50. Calculate the tension in the cable and the reaction at the hinge.

14 Two uniform rods each 3 m long and weighing 250 N are connected by a link bar (Fig. 4.51). Calculate the tensions in the link bar and chain neglecting their weights.

15 For each of the structures shown in Fig. 4.52(a), (b) and (c) determine the position of the resultant of the vertical loads; then, by using the triangle of forces, determine the tension in the cable and the reaction at the hinge. Neglect the weight of the rod.

Figure 4.51

Figure 4.52

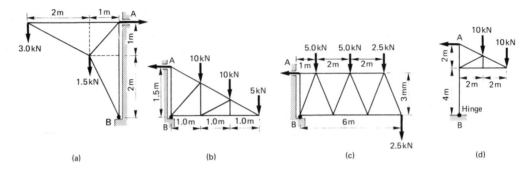

(a)   (b)   (c)   (d)

Figure 4.53

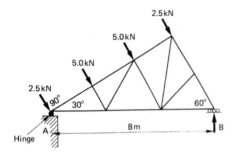

Figure 4.54

16 The weights of the structures shown in Fig. 4.53(a), (b), (c) and (d) may be neglected. In each case the reaction at A is horizontal, and B is a hinge. Determine the position of the resultant of the vertical loads; then, by applying the principle of the triangle of forces, determine the reactions at A and B.

17 The roof truss of Fig. 4.54 is supported on rollers at B and the reaction at B is therefore vertical. Determine the resultant of the four wind loads, then use the triangle of forces to determine the reactions at A and B.

18 Reduce the system of forces shown in Fig. 4.55 to a couple. Determine the arm and moment of the couple.

19 Calculate the reactions at A and B due to the given point loads in Fig.4.56(a) to (e). The loads are in kilonewtons (kN).

20 Calculate the reactions at A and B due to the given system of loads in Fig. 4.57(a) to (g).

21 A beam AB carries two point loads and is supported by two beams CD and EF as shown in plan in Fig. 4.58. Neglecting the weights of the beams, calculate the reactions at C, D, E and F.

22 A reinforced concrete bridge between two buildings spans 8 m. The cross-section of the bridge is shown in Fig. 4.59. The brickwork weighs 20 kN/m^3, the reinforced concrete weighs 23 kN/m^3, and the stone coping weighs 0.5 kN/m. The floor of the bridge has to carry a uniformly distributed load of 1.5 kN/m^2 in addition to its own weight. Calculate the force on the supporting buildings from one end of each beam.

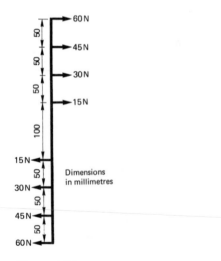

Dimensions in millimetres

Figure 4.55

Figure 4.56

Figure 4.57

Figure 4.58

Figure 4.59

5 Framed Structures

The framed structures dealt with in this chapter are pin-jointed, determinate, and both plane frameworks and three-dimensional or space frameworks.

The graphical solutions of plane frames by means of force diagrams are explained in some detail. The three analytical or calculation methods presented here are the Method of Sections, Resolution at Joints, and Tension Coefficients. Whilst the first two methods are applied to plane frames only, the usefulness of the application of the third one to space frames is also demonstrated.

A **frame** is a structure built up of three or more members which are normally considered as being pinned or hinged at the various joints. Any loads which are applied to the frame are usually transmitted to it at the joints, so that the individual members are in pure tension or compression.

A very simple frame is shown in Fig. 5.1. It consists of three individual members hinged at the ends to form a triangle, and the only applied loading consists of a vertical load of W at the apex. There are also, of course, reactions at the lower corners.

Under the action of the loads the frame tends to take the form shown in broken lines, i.e. the bottom joints move outward putting the member C in tension, and the members A and B in compression. Members A and B are termed **struts** and member C is termed a **tie.**

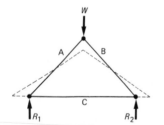

Figure 5.1

PERFECT, IMPERFECT AND REDUNDANT PIN-JOINTED FRAMES

A **perfect frame** is one which has just sufficient members to prevent the frame from being unstable.

Figure 5.2

Figure 5.3

Figure 5.4

Figure 5.5

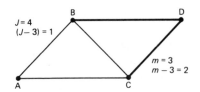

Figure 5.6

An **imperfect frame** is one which contains fewer than the number of members which would constitute a perfect frame.

Fig. 5.2 shows examples of perfect frames. Fig. 5.3 shows two examples of imperfect frames. In each of the two cases in Fig. 5.3 the frames would collapse as suggested by the broken lines; the addition of one member in each case would restore stability and produce perfect frames as in Fig. 5.2.

Fig. 5.4, for example, shows a perfect frame, and Fig. 5.5 (the same frame with the addition of member AB) a redundant frame.

Note: A redundant member is not necessarily a member having no load. The member AB (Fig. 5.5), for example, will be stressed, and will serve to create a stronger frame than that in Fig. 5.4, but the **redundant frame** (having one member more than is necessary to produce stability) cannot be solved by the ordinary methods of statics. It is, therefore, also called a **statically indeterminate** or hyperstatic frame.

Number of Members in a Perfect Frame

The simplest perfect frame consists of three members in the form of a triangle as in Fig. 5.6. The frame also has three joints or node points, A, B and C.

If *two* more members are added to form another triangle as in Fig. 5.7, then *one* more joint has been added.

So long as triangles only are added, then the frame will remain a perfect frame.

If J = number of joints, and m = number of members, then

$$2(J-3) = m - 3$$
$$2J - 6 = m - 3$$
$$2J = m - 3 + 6$$
$$2J = m + 3 \quad \text{or} \quad m = 2J - 3$$

i.e.

Number of members to form a perfect frame
= (twice the number of joints) − 3

or

$$\text{Number of joints} = \frac{\text{number of members} + 3}{2}$$

The frames shown in Fig. 5.8 all comply with this requirement.

The frames shown in Fig. 5.9 do not comply with this requirement, i.e. there are more than $2J - 3$ members, and the frames contain one or more redundant members.

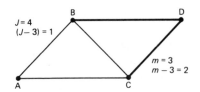

$J = 4$
$(J - 3) = 1$

$m = 3$
$m - 3 = 2$

Figure 5.7

Figure 5.8

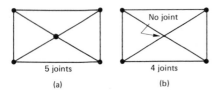

Figure 5.9

In Fig. 5.9(a) the number of members is 8, which is more than $(2 \times 5) - 3$. Similarly, in Fig. 5.9(b) the number of members is 6, which is greater than $(2 \times 4) - 3$.

GRAPHICAL SOLUTIONS FOR FRAMES: FORCE DIAGRAMS

Consider the simple perfect frame shown in Fig. 5.10. The known forces are the load at the apex and the two equal reactions of 10 kN.

In describing the loads, it is convenient to use Bow's Notation (see Chapter 2). Reading clockwise, the load at the apex, of 20 kN, is "load AB", the reaction at the right support is "load BC", and the reaction at the left support is "load CA". Similarly, the spaces *inside* the frame are usually denoted by numbers so that the force in the horizontal member is denoted by "force C1" or "force 1C".

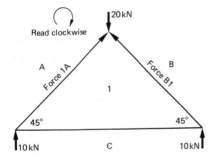

Figure 5.10

As explained in Chapter 2, the forces in the two inclined members can be obtained by considering the equilibrium of the joint at the apex (Fig. 5.11).

The force in the horizontal tie member can be obtained by considering either the joint at the left reaction (Fig. 5.12) or the joint at the right reaction (Fig. 5.13).

The three triangle of forces diagrams drawn separately for the three joints may, for convenience, be superimposed upon each other to form in one diagram the means of determining all unknown forces. This resultant diagram, which is called a **force diagram,** is shown in Fig. 5.14(b).

Note that no arrows are shown on the combined force diagram. The directions of the arrows in the frame diagram are obtained by considering each joint as in Figs. 5.11 to 5.13. One must consider each joint to be the centre of a clock, and the letters are read clockwise round this centre. Thus, having drawn the combined diagram, consider, say, the joint at the left reaction (Fig. 5.12). Reading clockwise round this joint, the inclined member is A1. On the force diagram,

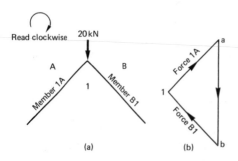

Figure 5.11 (*a*) Reading clockwise the forces are AB, B1 and 1A; (*b*) Force diagram (Note that the arrows follow each other round the triangle)

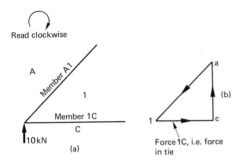

Read clockwise

Figure 5.12 Joint at left-hand reaction; (*b*) is the force diagram

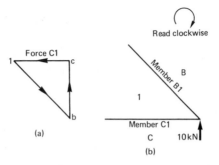

Read clockwise

Figure 5.13 Joint at right-hand reaction; (*a*) is the force diagram

Figure 5.14

the direction from *a* to 1 is downwards to the left, therefore the arrow is placed in the frame diagram as shown, near the joint. The tie is member 1C and 1 to *c* on the force diagram is a direction left to right, so the arrow is placed in this direction on the frame (or free body or space) diagram.

Considering the joint at the apex (Fig. 5.11), the left-hand inclined member is member 1A (not A1 as was the case when the joint at the left-hand reaction was being considered). From the force diagram 1 to *a* is upwards to the right and this fixes the direction of the arrow on the frame diagram. Similarly, for the same joint, the other inclined member is B1. From the force diagram *b* to 1 is upwards towards the left. (Study the force diagrams and the directions of the arrows in Figs. 5.11 to 5.13.)

Note the directions of the arrows indicating compression (strut) and tension (tie) respectively (Fig. 5.14(*a*)). These arrows represent the directions of the internal resistances in the members. Member A1 is a strut, which means that it has shortened as a result of the force in it. If the force were removed, the member would revert to its original length (i.e. it would lengthen) and the arrows indicate the attempt of the member to revert to its original length.

Member C1 is a tie and has therefore been stretched. The arrows indicate the attempt of the member to revert to its original (shorter) length.

The forces in the various members may be entered as shown in Table 5.1.

Table 5.1

| Member | Forces in kN | |
	Strut	Tie
A1	14	
B1	14	
C1		10

(a)

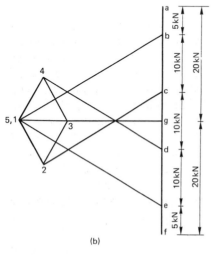

(b)

Figure 5.15 (*a*) Free-body, space or frame diagram; (*b*) Force diagram

Example 5.1 *Symmetrical Frame and Loading*

The principles described above may be applied to frames having any number of members. The work of determining the values of all the unknown forces may be enormously simplified by drawing *one* combined force diagram as shown in Fig. 5.15(*b*), and for this example the operations are explained in detail.

1 Starting from force AB (Fig. 5.15(*a*)) the *known* forces AB, BC, CD, DE, EF, FG and GA, working clockwise round the frame, are set down in order and to scale as *ab*, *bc*, *cd*, *de*, *ef*, *fg* and *ga* in Fig. 5.15(*b*): (Since the loading is symmetrical each of the reactions FG and GA is 20 kN.)

2 Point 1 on the force diagram is found by drawing *b*–1 parallel to B1, and *g*–1 parallel to G1. Point 1 lies at the intersection of these lines.

3 Point 2 on the force diagram is found by drawing *c*–2 parallel to C2 and 1–2 parallel to 1–2 from the frame (node 2). The point 2 is the intersection of these two lines.

4 Point 3 on the force diagram is found in the same manner by drawing 2–3 parallel to 2–3, and *g*–3 parallel to G3 from the frame, point 3 being the intersection.

5 The remaining points 4 and 5 are found in the same way as above, though in this case, as the frame is symmetrical and forces E5 and 4–5 are the same as B1 and 1–2, etc., only half the diagram need have been drawn.

6 The directions of the arrows may now be transferred to the frame diagram (Fig. 5.15(*a*)) by working clockwise around each node point.

Taking node 1 for example: moving clockwise; force AB is known to be downward, and *ab* (5 kN) is read downward on the force diagram. The next force in order (still moving clockwise round the joint) is force B1, which as *b*–1 on the force diagram acts to the left and downwards. Thus force B1 must act towards node 1 as indicated in Fig. 5.15(*a*). The next force in order (still moving clockwise) is 1G. 1–*g* on the force diagram acts from left to right, and thus the arrow is put in on the frame diagram acting to the right from node 1.

7 Node 2: BC is known to act downward. The next force (moving clockwise) is C2, which, as *c*–2 in the force diagram acts downward to the left, thus must act towards node 2. The arrow is shown accordingly. The next force (still moving clockwise) is 2–1, and as 2–1 on the force diagram acts upward to the left, this must act toward the node 2, as shown by the arrow.

8 When all nodes have been considered in this way, it will be seen that each member of the frame in Fig. 5.15(*a*) has two arrows, one at each end of the member. As explained earlier, these arrows indicate not what is being done to the

Table 5.2

Member		Forces in kN	
		Strut	Tie
B1	E5	30.0	
C2	D4	25.0	
G1	G5		26.0
	G3		17.5
1:2	4:5	9.0	
2:13	3:4		9.0

member, but what the member is doing to the nodes at each end of it.

Hence a member with arrows thus ←——→ is in compression, i.e. it is being compressed and is reacting by pressing outward.

Similarly, two arrows thus →←— indicate tension in the member.

9 The *amounts* of all the forces B1, C2, 1–2, etc., are now scaled from the force diagram, and the amounts and types of force can be tabulated as in Table 5.2.

It should be noted that constructing a force diagram involves considering each joint of the frame in turn and fixing the position of one figure on the force diagram by drawing two lines which intersect. No joint can be dealt with unless *all except two* of the forces meeting at the joint are known. This is why node 1 was taken as the first joint since, of the four forces meeting at this joint, only two B1 and 1G are unknown. Solving this joint gives the forces in the members B1 and 1G. Node 2 can now be dealt with since only two forces, C2 and 2–1, are unknown. Node 3 cannot be dealt with before node 2 because, although the force in the member G1 is known, there are still three unknown forces, 1–2, 2–3, and 3G.

Example 5.2 *Frame Containing a Member Having No Force*

The external loads and the two equal reactions are set off to scale as shown in Fig. 5.16(*b*).

Starting with node 1 the two unknown forces are A1 and 1E.

From *a* on the force diagram a vertical line must be drawn (parallel to A1) and point 1 must lie on this vertical line where it meets a horizontal line drawn parallel to member 1E. The point 1 must, therefore, be on the same point as *e* and force 1E = 0.

Considering node 2, point 2 is found by drawing 1–2 to intersect *b*–2.

Point 3 is found by drawing *c*–3 and 3–2 on the force diagram.

Since the frame and loading are symmetrical there is no need to draw further lines, but in this simple example, one more line, shown dotted, completes the force diagram.

Example 5.3 *Frame with Unsymmetrical Loading*

When the loading and/or the frame is unsymmetrical, *every* joint must be considered when drawing the force diagram.

Figure 5.16

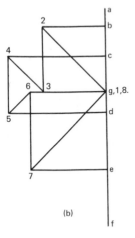

Figure 5.17

Fig. 5.17 is an example. [The reactions may be determined graphically using the link polygon (Chapter 3) or by calculation (Chapter 4).]

Example 5.4 *Frame with Loads Suspended from the Bottom Chord of the Frame in Addition to Loads on the Top Chord*

The reactions in Fig. 5.18 are determined in the usual manner and are 40 kN at the left support and 50 kN at the right support. The vertical load line is then drawn as follows: *a–b* is 20 kN down; *b–c* is 40 kN down. The next force is the reaction CD of 50 kN, therefore from *c* to *d* is measured upwards a distance equal to 50 kN. The next load is DE and is 20 kN down, therefore *d–e* represents this load on the force diagram. Similarly, *e–f* represents the load of 10 kN (EF) and finally *f–a* represents the vertical reaction FA of 40 kN.

The remaining force diagram can now be drawn in the usual manner, e.g. starting with node 1, two lines *a–1* and *f–1* are drawn to intersect at the point 1, and so on, as explained in Example 5.1.

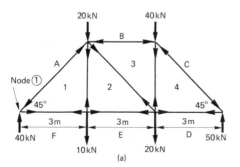

Example 5.5 *Frame Cantilevering over One Support*

Fig. 5.19 shows a frame cantilevering 2 m over the left support. By calculation, the reactions are found to be 7.5 kN and 4.0 kN respectively. The load line is drawn as explained in the previous example, *a–b, b–c, c–d, d–e* (all downwards), *e–f* upwards (4.0 kN), *f–g* downwards (1.5 kN), *g–a* upwards (7.5 kN).

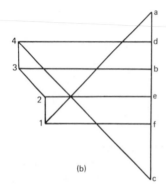

Figure 5.18

Example 5.6 *Cantilever Frames*

In Fig. 5.20 there is no need to determine the reactions before starting on the force diagram. The loads are first plotted in order, *a–b, b–c, c–d, d–e*. Starting with node 1, *b–1* and *1–a* fix the position of point 1 on the force diagram, and so on, finishing with node 7, i.e. 6–7 and 7–a. Since member 7A is in compression, the reaction at the bottom support will be equal in value to the force in the member, therefore *f* is at the same point as 7. The reaction at the top is given by the line *e–f* on the force diagram, whilst line *a–f* represents the reaction at the bottom.

(a)

(b)

Figure 5.19

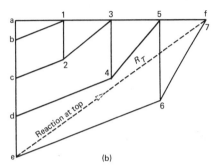

(b)

Figure 5.20

Frames Subjected to Inclined Loads

The frames dealt with in the preceding examples have all carried purely vertical loads, and so the end reactions have also been vertical.

Sometimes roof trusses have to be designed to withstand the effect of wind, and this produces loads which are assumed to be applied to the truss (via the purlins) at the panel points, in such a way that the panel loads are acting normal to the rafter line. As the inclined loads have a horizontal thrust effect on the truss as well as a vertical thrust effect, it follows that at least one of the reactions will have to resist the horizontal thrust effect.

It was stated in Chapter 3 that, when a truss is subjected to inclined loads, the usual assumptions are either that *a*) one of the reactions is vertical, or *b*) the horizontal components of the two reactions are equal. The method of determining such reactions by the use of the link polygon has been described in Chapter 3.

When only symmetrical inclined loads have to be considered, and one of the reactions is assumed to be vertical as in Fig. 5.21, the reactions can be found more easily by the method described below than by the link polygon.

Figure 5.21

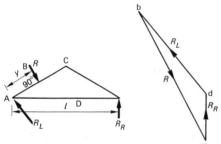

Figure 5.22

Table 5.3

| Member | Forces in kN | |
	Strut	Tie
B1	7.0	
C2	7.0	
D4	5.8	
D5	5.8	
E1		9.8
E3		5.0
E5		5.0
1–2	5.0	
2–3		5.0

No force in 3–4 and 4–5

If the four loads are considered to be replaced temporarily by the resultant R, then only three loads act on the truss, i.e. R, R_L and R_R. The three loads must act through one point for equilibrium, and the loads R and R_R obviously act through point x, the point of intersection of the vertical reaction R_R and the resultant inclined load R. As R_L must act through point x, the direction of R_L is found by joining point A to point x.

The direction of all three loads is thus known, together with the magnitude of the resultant R, so the amounts of R_L and R_R may be found graphically as in Fig. 5.21.

The resultant R is set down to scale (b to c). The point d is found by drawing in the direction of R_R from c and the direction of R_L from b to their intersection point d.

This vector diagram bcd is, of course, the triangle of forces for the loads, and cd = reaction R_R to scale, and db = reaction R_L to scale. From these three points, b, c, and d, the force diagram for the frame can be drawn in the usual way.

The vertical reaction can also be calculated by taking moments about the other end of the truss. For example, from Fig. 5.22, taking moments about A

$$R_R \times L = R \times y$$

$$R_R = \frac{R \times y}{L}$$

Point d can now be found by setting down the resultant load R (line bc) followed by the vertical reaction R_R as calculated

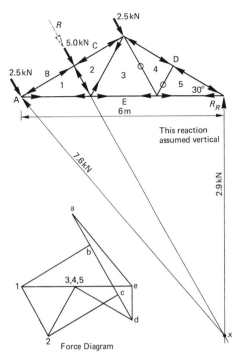

Figure 5.23

to give the point d. Reaction R_L will then be found by scaling force db.

Example 5.7 *Frame with Inclined Loads*

When the two reactions have been found, either graphically or by calculation, the forces in the individual bars may be found by drawing a force diagram, as in Fig. 5.23.

The resultant load R (inclined) is produced to intersect the vertical reaction R_R at the point x. The direction of left-hand reaction R_L is then found by joining x to A.

The two reactions and the the three inclined loads are thus now all known, and may be set down as in Fig. 5.23 in the order ab, bc, cd (inclined), de (vertical), and ea parallel to R_L.

The points 1, 2, 3, 4, 5 on the force diagram are then found in the usual way, as shown in Fig. 5.23.

Note that there is no force in members 3–4 and 4–5 from this system of loading (see Table 5.3).

Another example is shown in Fig. 5.24 and Table 5.4.

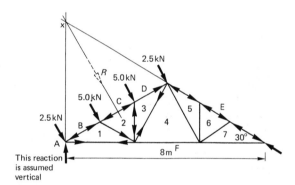

Table 5.4

| Member | Forces in kN | |
	Strut	Tie
B1	12.7	
C2	10.0	
D3	13.1	
E5	8.7	
E6	8.7	
E7	8.7	
F1		9.7
1–2	5.7	
2–3	6.0	
3–4		10.0

No force in 4–5, 5–6, 6–7, F4 and F7

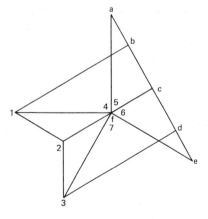

Force Diagram

Figure 5.24

CALCULATION METHODS FOR FRAMES

When the forces in *all* the members have to be found, the force diagram is usually the most convenient method, but on occasions it is necessary to determine the force in a single individual member. In cases of this type it is normally more convenient to calculate the amount of the force.

Method of Sections

Consider the N girder shown in Fig. 5.25. It is required to calculate the force in the member 1–2 of the bottom boom.

1 Imagine the girder to be cut completely through, along the section S–S, a section passing through the member 1–2 concerned and two other members, 3–4 and 3–2.

2 Consider the portion of truss to the right of line S–S to be removed. The portion to the left of line S–S would then, of course, collapse, because three forces (the forces in members 1–2, 3–2 and 3–4) which were necessary to retain equilibrium had been removed.

3 If now, three forces *x*, *y* and *z*, equal respectively to force 1–2, force 3–4 and force 3–2, are now applied to the portion of frame concerned, as shown in Fig. 5.26, then the portion of frame will remain in equilibrium under the action of the 50 kN reaction, the two 20 kN loads, and the forces *x*, *y* and *z*.

4 These three forces (*x*, *y* and *z*) are as yet unknown in amount and direction, but the force in member 1–2, i.e. force *x*, is to be determined at this stage, and it will be seen that the other two unknown forces (*y* and *z*) intersect at and pass through the point 3 so that they have no moment about that point.

5 Thus, taking moments about the point 3, the portion of frame to the left of S–S is in equilibrium under the moments of the two applied loads, the moment of the reaction, and the moment of the force *x*.

The moment of the reaction is a clockwise one of (50×3), i.e. 150 kN m.

The moment of the applied load is anticlockwise about point 3 and equals (20×3), i.e. 60 kN m.

For equilibrium the moment of force *x* about point 3 (i.e. $3x$) must equal $(150-60)$, i.e. 90 kN m anticlockwise.

Therefore $3x = 90$, whence $x = 30$ kN and the arrow must act in the direction shown in Fig. 5.26 (in order that the moment of force *x* about point 3 is anticlockwise).

The member 1–2 is therefore in tension (pulling away from joint 1).

Figure 5.25

Figure 5.26

Note: It will be seen that, in general, the rule must be: take a cut through three members (which include the one whose force is to be determined) and take moments about the point through which the lines of action of the other two forces intersect.

To determine the force in the member 3–4:
Take moments about the point 2 where forces x and z intersect.

Moment of reaction $= (50 \times 6)$
$$= 300 \text{ kN m} \quad \text{clockwise}$$

Moments of applied loads $= (20 \times 6) + (20 \times 3)$
$$= 180 \text{ kN m} \quad \text{anticlockwise}$$

Therefore

Moment of force $y = 3y = 300 - 180$
$$= 120 \text{ kN m} \quad \text{anticlockwise}$$
$$y = 40 \text{ kN}$$

and, for the moment to be anticlockwise about point 2, the arrow must act towards the joint 3, therefore member 3–4 is a strut.

It should be noted that this particular method has its limitations. It would not appear to apply, for example, to member 3–2 (force z) for no point exists through which forces x and y intersect, as x and y are, in fact, parallel to each other.

However, when force y has been calculated, the force z may be found by taking moments about point 1, including the known value of force y (member 3–4).

Note: force y is 40 kN and is anticlockwise about point 1.

Moment of the reaction about point 1
$$= (50 \times 3) = 150 \text{ kN m}$$

Moments of the applied loads and the force in member 3–4
$$= (20 \times 3) + (40 \times 3) = 180 \text{ kN m}$$

Therefore

Moment of force z about point 1
$$= 180 - 150 = 30 \text{ kN m} \quad \text{clockwise}$$

Hence

$$2.1z = 30$$
$$z = 14.2 \text{ kN}$$

The arrow must act as shown in Fig. 5.26, therefore member 3–2 is a tie.

Figure 5.27

It should be noted that these arrows must be considered in relation to the nearest joints of that portion of the frame which remains after the imaginary cut through the three members has been made. These results should be checked by drawing a force diagram for the entire frame.

Example 5.8 *Roof Truss with Vertical Loads*

Calculate the forces in the members marked 2–3, 1–6 and 2–6 of the frame shown in Fig. 5.27.

Solution Member 2–3: take moments about point 6 (Section at S–S).

Moment of reaction $= 12 \times 4$

$$= 48 \text{ kN m} \quad \text{clockwise}$$

Moment of applied loads $= (3 \times 4) + (6 \times 2)$

$$= 24 \text{ kN m} \quad \text{anticlockwise}$$

Moment of force in member 2–3 $= 48 - 24$

$$= 24 \text{ kN m} \quad \text{anticlockwise}$$

Therefore

Force 2–3 $\times 2$ m $= 24$ kN m

Force 2–3 $= 12$ kN   (strut)

Member 1–6: take moments about point 2

Moment of reaction $= 12 \times 2$

$$= 24 \text{ kN m} \quad \text{clockwise}$$

Moment of loads $= 3 \times 2$

$$= 6 \text{ kN m} \quad \text{anticlockwise}$$

Force 1–6 $\times 1.15$ m (anticlockwise) $= 24 - 6$

Force 1–6 $= 18/1.15$

$$= 15.7 \text{ kN} \quad \text{(tie)}$$

Member 2–6: take moments about point 1

Moment of reaction $= 0$

Moment of loads $= 6 \times 2$

$$= 12 \text{ kN m} \quad \text{clockwise}$$

Force 2–6 $\times 2$ (anticlockwise) $= 12$ kN m

Force 2–6 $= 6$ kN   (strut)

Remember that, for determining whether members are struts or ties, the arrows as shown in Fig. 5.27 must be considered as acting towards or away from the nearest joint in that portion of the frame which remains after the *cut* has been made.

Figure 5.28

Figure 5.29

Figure 5.30

Example 5.9 *Girder with Parallel Flanges*

The application of this method may be simplified considerably when the frame concerned has parallel flanges. Consider, for example, the Warren girder shown in Fig. 5.28.

In using the method of sections to find the force in member EF, the cut would be made through the section X–X, and moments taken about the point C.

Taking moments about C to the right,

$$(\text{Force in EF}) \times h = \text{moments of forces } R_R \text{ and } W_3$$
$$= (R_R \times x_2) - (W_3 \times x_1)$$
$$\text{Force in EF} = \frac{(R_R \times x_2) - (W_3 \times x_1)}{h}$$

But the value $(R_R \times x_2) - (W_3 \times x_1)$ is obviously the bending moment at C, considering the girder as an ordinary beam as shown in Fig. 5.29.

So the *rule* may be given as follows:

To calculate the force in any top or bottom boom member of a girder of this type, calculate the bending moment at the node point opposite to the member and divide by the vertical height of the girder, e.g.

$$\text{Force in EF} = \text{BM at point C} \div h$$
$$\text{Force in DC} = \text{BM at point E} \div h$$
$$\text{Force in GE} = \text{BM at point D} \div h$$

Note: This simplification of the method is sometimes known as *the method of moments*.

Example 5.10

In Fig. 5.30, by calculation, the reactions are:

$$R_L = 45 \text{ kN} \quad \text{and} \quad R_R = 55 \text{ kN}$$

BM at B and G $= (45 \times 3) = 135$ kN m

BM at C and H $= (45 \times 6) - (20 \times 3) = 210$ kN m

BM at D and J $= (55 \times 3) = 165$ kN m

Therefore, considering members cut by Section 1,

$$\text{Force in AB} = \frac{\text{BM at G}}{h} = 135/3 = 45 \text{ kN}$$

because the two other members meet at G.

$$\text{Force in FG} = \frac{\text{BM at A}}{h} = 0/3 = 0$$

because the two other members meet at A.

Next, consider members cut by Section 2.

$$\text{Force in BC} = \frac{\text{BM at H}}{h} = 210/3 = 70 \text{ kN}$$

because the other two members meet at H

$$\text{Force in GH} = \frac{\text{BM at B}}{h} = 135/3 = 45 \text{ kN}$$

because the other two members meet at B.
Hence

$$\text{Force in CD} = \frac{\text{BM at H}}{h} = 210/3 = 70 \text{ kN}$$

$$\text{Force in HJ} = \frac{\text{BM at D}}{h} = 165/3 = 55 \text{ kN}$$

$$\text{Force in DE} = \frac{\text{BM at J}}{h} = 165/3 = 55 \text{ kN}$$

$$\text{Force in JK} = \frac{\text{BM at E}}{h} = 0/3 = 0.$$

Method of Resolution of Forces at Joints

The forces in the individual members of loaded frames may also be determined by considering the various forces acting at each node point. This method is particularly useful for dealing with those members not so easily tackled by the method of sections, e.g. the vertical members of the frame dealt with in the previous example.

Example 5.11

Consider the node *F* in Fig. 5.31.

There is a load of 45 kN *upward* at F. Obviously, member AF must itself counteract this (as FG, a horizontal member, cannot resist vertical force). Hence the force in member AF must equal 45 kN, and it must, at the node point F, be acting downwards in opposition to the reaction R_L, as shown by the arrowhead 1. This shows member AF to be a strut, and the other arrowhead 2 may be put in, acting towards A.

Consider the node A.

It is now known that a force of 45 kN acts upwards at this point (arrowhead 2). Again, member AB, being horizontal, cannot resist a vertical load, so member AG must resist the upward 45 kN at A. Thus the force in AG must act downwards at A, and the arrow 3 may be placed in position. The amount of force AG must be such that its vertical component equals 45 kN as in Fig. 5.32. Thus

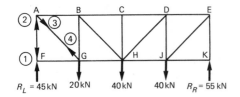

$R_L = 45$ kN   20 kN   40 kN   40 kN $R_R = 55$ kN

Figure 5.31

Figure 5.32

Figure 5.33

Figure 5.34

$$\frac{45}{\text{force AG}} = \sin 45° = 0.7071$$

Therefore

Force in AG $= 45/0.7071 = 63.6$ kN

and AG is in tension, so the arrow 4 may be drawn in.

The load of 63.6 kN acting down and to the right at A tends to pull point A to the right, horizontally. Thus the force in member AB at A must *push* point A to the left for equilibrium (arrow 5, showing member AB to be a strut); thus arrow 6 may be placed as shown in Fig. 5.33. The amount of force in AB is such that it counteracts the horizontal component of force 63.6 kN (arrow 3).

Thus, as in Fig. 5.34,

$$\frac{\text{Force in AB}}{63.6} = \cos 45° = 0.7071$$

Thus force AB $= 63.6 \times 0.7071 = 45$ kN compression, as was seen by the method of moments.

Consider the node G.

It has already been seen that there is no force in member FG, since the bending moment is zero at the opposite point A. Thus, as there is a force of 63.6 kN at G, acting upwards and to the left, the force in BG must act in such a way that it counteracts the resultant of the vertical forces at G. These vertical forces are

  a) the vertical resultant upward of the 63.6 kN
      (i.e. 45 kN ↑);
  b) the downward applied load of 20 kN ↓.

Their resultant is obviously 25 kN upward, so member BG must act downwards to the extent of 25 kN (arrows 7 and 8 may thus be placed) and BG is thus a strut with a force of 25 kN (Fig. 5.35). Also, as AG pulls to the *left* at G to the extent of 45 kN (the horizontal component of the force AG) then member GH must pull to the *right* to the extent of 45 kN (arrow 9), and member GH is seen to be a tie with a force of 45 kN. This again agrees with the result obtained by the method of moments.

This process may be continued from each node to the next, and although its application is admittedly somewhat more laborious than the drawing of a force diagram, this method is strongly recommended. Its use most definitely results in a better and more complete understanding of the forces acting at each node, and by its use struts and ties are more easily distinguished.

Figure 5.35

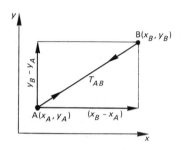

Figure 5.36

Method of Tension Coefficients

This is a modified version of the previous method and it is particularly useful in the case of pin-jointed, three-dimensional or space frames.

The basis of this method is the resolution into components of the external and internal forces acting on each joint of the frame, using the lengths of members and the coordinates of joints. All the members of the frame are initially assumed to be in tension.

Fig. 5.36 shows bar AB, a member of a pin-jointed frame, assumed to be subject to a tensile force T_{AB}.

Let the length of bar AB be L_{AB} and the coordinates of joint A be x_A, y_A and those of joint B, x_B, y_B. Then, resolving the force T_{AB} in the x direction, gives

$$T_{AB} \times \frac{x_B - x_A}{L_{AB}} = \frac{T_{AB}}{L_{AB}} \times (x_B - x_A)$$

In this expression the factor T_{AB}/L_{AB}, i.e. tension in bar AB divided by the length of the bar, is called the **tension coefficient** for the bar and is denoted by the symbol t_{AB}.

Hence the component in the x direction is

$$t_{AB}(x_B - x_A)$$

and, similarly, the component in the y direction is

$$t_{AB}(y_B - y_A)$$

Where there are a number of members and applied loads acting at a joint, which is in equilibrium under the action of the internal and external forces, two sets of equations can be formed, one for each of the directions x and y.

Example 5.12

A load of 7.2 kN is suspended from a soffit by two ropes PQ and QR (Fig. 5.37). Determine the forces in the ropes.

Solution The equations for joint Q (i.e. assuming the origin of coordinates is at Q) in terms of tension coefficients are:

(1) in direction x $\quad -1.8t_{PQ} + 3.2t_{QR} = 0$

(2) in direction y $\quad +2.4t_{PQ} + 2.4t_{QR} - 7.2 = 0$

Then from (2)

$$t_{PQ} = 3.0 - t_{QR}$$

and substituting into (1)

$$-1.8(3.0 - t_{QR}) + 3.2t_{QR} = 0$$
$$t_{QR} = 5.4/5.0 = 1.08$$

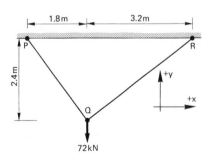

Figure 5.37

and

$$t_{PQ} = 3.0 - 1.08 = 1.92$$

But $t_{PQ} = T_{PQ}/L_{PQ}$ and thus $T_{PQ} = t_{PQ} \times L_{PQ}$. Therefore

$$T_{PQ} = 1.92 \times \sqrt{[(2.4)^2 + (1.8)^2]}$$
$$= 1.92 \times 3.0 = 5.76 \text{ kN}$$

and

$$T_{QR} = 1.08 \times \sqrt{[(2.4)^2 + (3.2)^2]}$$
$$= 1.08 \times 4.0 = 4.32 \text{ kN}$$

It should be noted that both T_{PQ} and T_{QR} have positive values and, therefore, they are ties. A negative solution denotes a compression member, i.e. a strut.

From the above example it follows that, for a frame having J joints, $2J$ equations can be formed. This, incidentally, is the same number of equations as that used in the Method of Resolution at Joints, which may explain the lack of popularity of the Method of Tension Coefficients for plane frames. It is, however, very useful for computer application, particularly in the solution of space frames.

A space frame has to be considered in three dimensions, hence the components of the forces at each joint are resolved in three directions x, y and z. Consequently, there will be $3J$ equations for the frame and the number of members in a perfect pin-jointed space frame is $m = 3J - 6$. Most space frames, however, will have pinned supports, so the number of members in such a frame is $m = 3J_f$, where J_f denotes a free joint (i.e. a joint other than at a support).

(a) Plan

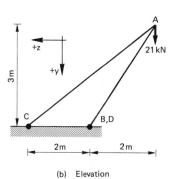

(b) Elevation

Figure 5.38

Example 5.13

The three equations for joint A of the shear legs shown in Fig. 5.38 are formed as follows:

(1) in direction x $+2t_{AB} - 2t_{AD} = 0$
(2) in direction y $+3t_{AB} + 3t_{AC} + 3t_{AD} + 21 = 0$
(3) in direction z $+2t_{AB} + 4t_{AC} + 2t_{AD} = 0$

Then from (1), $t_{AB} = t_{AD}$, and adding (1) and (3) gives

$$t_{AB} = -t_{AC}$$

and substituting both into (2),

$$-3t_{AC} + 3t_{AC} - 3t_{AC} = -21$$
$$t_{AC} = 7 \quad \text{and} \quad t_{AB} = -7 = t_{AD}$$

Therefore

$$T_{AB} = -7 \times \sqrt{[2^2 + 2^2 + 3^2]}$$
$$= 28.86 \text{ kN} \quad \text{strut}$$
$$T_{AC} = +7 \times \sqrt{[4^2 + 3^2]}$$
$$= 35.00 \text{ kN} \quad \text{tie}$$
$$T_{AD} = 28.86 \text{ kN} \quad \text{strut}$$

Example 5.14

In the case of Fig. 5.39 the simple space frame is supporting an inclined load. It is, therefore, necessary to resolve the load into components in directions x, y and z, before writing down the usual equations.

Let the load be F and its components F_x, F_y and F_z, respectively. Therefore

$$F_x = 0$$
$$F_y = 3 \times \cos 30° = 2.6 \text{ kN}$$
$$F_z = 3 \times \sin 30° = 1.5 \text{ kN}$$

It is usually helpful, particularly for a many-jointed frame, to present the equations and the resulting tension coefficients in tabular form as shown in Table 5.5.

Figure 5.39

(a) Plan

(b) Elevation

Hence

$$T_{AB} = T_{AD} = -0.49 \times \sqrt{[(0.7)^2 + (1.5)^2]}$$
$$= -0.811 \text{ kN} \quad \text{(negative, i.e. strut)}$$

and

$$T_{AC} = -0.75 \times \sqrt{[(2.0)^2 + (1.5)^2]}$$
$$= -1.875 \text{ kN} \quad \text{(also a strut)}$$

Table 5.5

Joint	Direction	Equation	Tension Coefficient
A	x	$-0.7t_{AB} + 0.7t_{AD} = 0$	$t_{AB} = t_{AD}$
	y	$+2.6 + 1.5t_{AB} + 1.5t_{AC} + 1.5t_{AD} = 0$	$t_{AB} = -0.49$
	z	$1.5 + 2.0t_{AC} = 0$	$t_{AC} = -0.75$

EXERCISE 5

1–15 Determine the reactions and the type and magnitude of the forces in the members of the frames shown in Figs. 5.40–5.54. Show the results in tabular form.

Figure 5.40

Figure 5.41

Figure 5.42

Figure 5.43

Figure 5.44

Figure 5.45

Figure 5.46

Figure 5.47

Figure 5.48

Figure 5.49

Figure 5.50

Figure 5.51

Figure 5.52

Figure 5.53

Figure 5.54

16–18 For the frames shown in Figs. 5.55–5.57 determine:
i) the directions and values of the reactions;
ii) the type and magnitude of the forces in all the members.

19 A framed structure is supported at A and B in such a manner that the reaction at A is horizontal. Calculate the values of the two reactions and determine the type and magnitude of the forces in the members of that structure shown in Fig. 5.58.

Figure 5.55

Figure 5.58

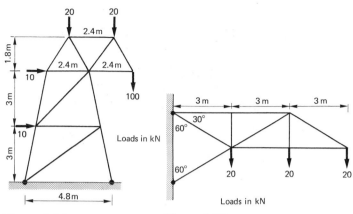

Figure 5.56

Figure 5.57

20–23 The right-hand ends of the frames shown in Figs. 5.59–5.62 are supported on rollers, i.e. the reaction there is vertical as indicated. Determine the reactions and the type and magnitude of the forces in the members of the frames.

Figure 5.59

Figure 5.60

Figure 5.61

Figure 5.62

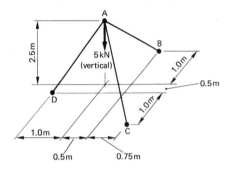

Figure 5.63

24–27 Repeat Problems 20–23 assuming that the roller supports are now at the left-hand ends of the frames.

28–31 Determine the reactions and the type and magnitude of the forces in the members in the frames shown in Figs. 5.59–5.62, assuming that the two supports resist equally the horizontal components of the inclined loads.

32–34 Using the Method of Tension Coefficients determine the forces in the space frames shown in Figs. 5.63–5.65.

Figure 5.64

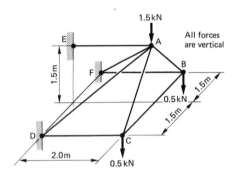

Figure 5.65

6 Stress, Strain and Elasticity

Stress, strain and elasticity are dealt with in the context of direct tensile and compressive stresses and their relationship is explained in terms of the Modulus of Elasticity. The ultimate and yield stresses are considered, and the need for a reduced working or permissible stress and the idea of a factor of safety are identified. The modular ratio for composite elements is introduced.

STRESS

When a member is subjected to a load of any type, the many fibres or particles, of which the member is made up, transmit the load throughout the length and section of the member, and the fibres doing this work are said to be in a state of stress.

There are different types of stress, but the principal kinds are tensile and compressive stresses.

Tensile Stress

Consider the steel bar shown in Fig. 6.1 having a cross-sectional area of A mm^2 and pulled out at each end by forces W. Note that the total force in the member is W kN (not

Figure 6.1

X

W kN

Stress resisting tearing
of the fibres

X

W kN

Area of cross-section
= A mm^2

2W kN). There would be no force at all in the member if only one load W was present as the member would not be in equilibrium.

At *any* plane such as X–X taken across the section, there exists a state of stress between the fibres on one side of the plane and those on the other. Here the stress is tensile by nature—the fibres on one side exerting stress to resist the tendency of being pulled away from the fibres on the other side of the plane. This type of resistance to the external loads is set up along the whole length of the bar and not at one plane only, just as in a chain resisting forces are set up by every link.

In this particular case (Fig. 6.1), if the loads act through the centroid of the shape, the stress is provided equally by the many fibres and each square millimetre of cross-section provides the same resistance to the "pulling apart" tendency.

This is known as **direct** or **axial stress.**

Fig. 6.2 represents a bar of cast steel which is thinner at the middle of its length than elsewhere, and which is subjected to an axial pull of 45 kN. If the bar were to fail in tension, it would be due to the bar snapping where the amount of material is a minimum. The total force tending to cause the bar to fracture is 45 kN at all cross-sections, but whereas 45 kN is being resisted by a cross-sectional area of 1200 mm² for part of its length it is being resisted by only 300 mm² at the middle portion of the bar. The intensity of load is greatest at this middle section and is at the rate of 150 N/mm² of cross-section. At other points along the bar, 45 kN is resisted by 1200 mm² and the stress is equal to 37.5 N/mm².

In cases of *direct* tension, therefore,

$$\text{Stress} = \frac{\text{Applied load}}{\text{Area of cross-section of member}} = \frac{W}{A}$$

The SI unit for stress is the pascal, $\text{Pa} = 1\ \text{N/m}^2$. However, since the square metre is a rather inconvenient unit of area when measuring stress, British Standards have adopted the multiple N/mm² as a stress unit for materials.

It should be noted that $1\ \text{N/mm}^2 = 1\ \text{MN/m}^2 = 1\ \text{MPa}$.

Example 6.1

A bar of steel 2000 mm² in cross-sectional area is being pulled with an axial force of 180 kN. Find the stress in the steel.

Solution Since 2000 mm² of cross-section is resisting 180 kN or 180 000 N it means that each mm² is resisting 90 N. In other words,

$$\text{Stress} = \frac{W}{A} = \frac{180\ 000\ \text{N}}{2000\ \text{mm}^2} = 90\ \text{N/mm}^2$$

Figure 6.2

45 kN

12 mm

24 mm

50 mm 25 mm

Area of cross-section = 300 mm²

Area of cross-section = 1200 mm²

45 kN

Figure 6.3

Compressive Stress

Fig. 6.3 shows a similar member to that of Fig. 6.1 but with two axial forces of W kN each acting inwards towards each other and thus putting the bar into a state of compression.

Again, at any plane section such as X–X there is a state of stress between the fibres, but this time the stress which is generated is resisting the tendency of the fibres to be crushed. Once more the stress is shared equally between the fibres, and the stress is

$$\frac{\text{Load}}{\text{Area}} \quad \text{i.e.} \quad \frac{W}{A} \text{ MN/m}^2 \text{ or N/mm}^2$$

Example 6.2

A brick pier is 0.7 m square and 3 m high and weighs 19 kN/m^3. It is supporting an axial load from a column of 490 kN (Fig. 6.4). The load is uniformly spread over the top of the pier so the arrow shown merely represents the resultant of the load. Calculate *a*) the stress in the brickwork immediately under the column; *b*) the stress at the bottom of the pier.

Solution a) Area of cross-section = 0.49 m^2. 490 kN on 0.49 m^2 is equivalent to 1000 kN on each square metre, or

Stress = 490 kN/0.49 m^2 = 1 MN/m^2 or 1 N/mm^2

b) Weight of pier = $0.7 \times 0.7 \times 3.0 \times 19 = 28$ kN, hence

Total load = $490 + 28 = 518$ kN

and

Stress = 518 kN/0.49 m^2 = 1057 kN/m^2 or 1.06 N/mm^2

Figure 6.4

STRAIN

All materials alter slightly in their shape when they are stressed. A member which is subjected to tensile stress increases in length and its cross-section becomes slightly smaller. Similarly, a compression member becomes shorter and slightly larger in cross-section. The very slight alteration in cross-section of building members is not as a rule important and only the alterations in length will be considered in this book.

When a tension force is applied to a rubber band it becomes longer, and when the force is removed, the material reverts to its original length. This property is common to some building materials such as steel, although under its normal working loads the amount of elongation is usually too small to be detected by the unaided eye. Another property of elastic materials is that the alteration in length is directly proportional to the load. For example, if in a given member

10 kN produces an extension of 2 mm
20 kN will produce an extension of 4 mm
30 kN will produce an extension of 6 mm

and so on, provided the *elastic limit* of the material is not exceeded.

This law

change in length is proportional to the force

was first stated by Robert Hooke (1635–1703) and is therefore known as **Hooke's Law.** Hooke was a mathematician and scientist and a contemporary of Sir Christopher Wren.

Although all materials alter in length when stressed, materials which are not elastic do not obey Hooke's Law. That is, the changes in length are not directly proportional to the load. Structural steel and timber are almost perfect elastic materials.

Just as it is convenient to represent the stress in a member as force per unit area, so it is convenient to represent the change in length of a member in terms of change per unit length, i.e.

$$\text{Strain} = \frac{\text{Change in length}}{\text{Original length}} = \frac{\delta l}{l}$$

where δ is a Greek symbol called delta which is frequently used to denote small changes.

Since strain is the direct outcome of stress it also is classified as tensile and compressive.

ELASTICITY

Suppose that a general formula is required by the use of which the amount of elongation (or shortening) in any member composed of an elastic material can be determined. Let that material be rubber. By carrying out experiments on pieces of rubber of different lengths, different areas of cross-section and different qualities, the following facts would be demonstrated:

1 Increase in the load W produces proportionate increase of elongation (Hooke's Law), therefore W is one term of the required formula.

2 Increase in length l of the member means increase in elongation. It will be found that the elongation is directly proportional to the length of the member, i.e. for a given load and a given area of cross-section a member 500 mm long will stretch twice as much as a member of the same material 250 mm long. l is therefore a term in the formula.

3 It is harder to stretch a member of large cross-sectional area A than a member of small cross-sectional area. Experiments will show that the elongation is inversely proportional to the area, i.e. for a given load and a given length a member 10 mm^2 in cross-sectional area will stretch only half as much as a member 5 mm^2 in area. A is therefore a term in the formula.

4 Rubber can be obtained of different qualities, i.e. of different stiffnesses. Other conditions being constant (i.e. length, area, etc.) it is more difficult to stretch a "stiff" rubber than a more flexible rubber. Let this quality of stiffness be represented by the symbol E, then the greater the value of E, i.e. the greater the stiffness of the material, the smaller will be the elongation; other conditions (length, load and area) being constant.

The formula for elongation (or shortening) of an elastic material can now be expressed as

$$\delta l = \frac{Wl}{AE}$$

W and l are in the numerator of the fraction, because the greater their values, the greater the elongation. A and E are in the denominator of the fraction because the greater their values (i.e. the greater the area and the greater the stiffness) the less will be the elongation.

The symbol E in the above formula represents the stiffness of a given material, i.e. it is a measure of its elasticity and is called the **modulus of elasticity**. If it can be expressed in certain definite units, the formula can be used to determine elongations of members of structures. Now if $\delta l = Wl/AE$, by transposing the formula,

$$E = Wl/A\,\delta l$$

and the value of E for different materials can be obtained experimentally by compression or tension tests.

The foregoing explanation is lengthy and the usual more concise treatment of the subject is as follows. In an elastic material stress is proportional to strain (Hooke's Law), i.e.

$$\frac{\text{Stress}}{\text{Strain}} = \text{a constant value}$$

This value is called the modulus (i.e. measure) of elasticity of the material and is denoted by E.

$$\text{Therefore} \quad \frac{\text{Stress}}{\text{Strain}} = E$$

$$\text{but} \quad \text{Stress} = \frac{\text{load}}{\text{area}} = \frac{W}{A}$$

$$\text{and} \quad \text{Strain} = \frac{\text{change in length}}{\text{original length}} = \frac{\delta l}{l}$$

Thus

$$\frac{W}{A} \Big/ \frac{\delta l}{l} = E \quad \text{or} \quad E = \frac{Wl}{A\,\delta l}$$

The modulus of elasticity is often called Young's modulus, after the scientist Thomas Young (1773–1829). Note that modulus of elasticity is expressed in the same units as stress. This is because stress is load per unit area and strain has no dimensions, it is just a number and therefore

$$\frac{\text{Stress}}{\text{Strain}} \left(\frac{\text{MN/m}^2 \text{ or N/mm}^2}{\text{a number}} \right) = E \text{ N/mm}^2$$

Fig. 6.5 indicates in essentials the manner of testing a block of timber 75 mm × 100 mm in cross-section and 300 mm high. In the actual experiment four gauges (extensometers) were used, one at each corner, and the average of the four readings was taken for each increment of load. A gauge was used at each corner because possible small inequalities in the top and bottom surfaces of the timber might result in unequal shortening of the timber block. The following readings (Table 6.1), which have been slightly adjusted for purposes of clearer explanation, were recorded as the compressive load was gradually increased.

Top plate of compression machine

Dial gauge or extensometer

300 mm

Bottom plate

Figure 6.5

Figure 6.6

Now $E = \dfrac{Wl}{A\,\delta l}$, where $l = 300\text{ mm}$ and $A = 7500\text{ mm}^2$.

Hence

from reading 2 $\quad E = \dfrac{20\,000 \times 300}{7500 \times 0.1} = 8000\text{ N/mm}^2$

from reading 3 $\quad E = \dfrac{40\,000 \times 300}{7500 \times 0.2} = 8000\text{ N/mm}^2$

from reading 4 $\quad E = \dfrac{60\,000 \times 300}{7500 \times 0.3} = 800\text{ N/mm}^2$

from reading 5 $\quad E = \dfrac{80\,000 \times 300}{7500 \times 0.4} = 8000\text{ N/mm}^2$

Table 6.1

	Load on specimen W	Shortening of specimen δl
1	0 kN	0.0 mm
2	20 kN	0.1 mm
3	40 kN	0.2 mm
4	60 kN	0.3 mm
5	80 kN	0.4 mm

The results of the experiment can be plotted as shown in Fig. 6.6. In an actual experiment results might not be as uniform as given in Table 6.1, but the resultant graph would approximate very closely to a straight line.

This indicates that E is a constant value, and it is this value which is taken as the modulus of elasticity for the timber.

Example 6.3

A post of timber similar to that used in the above test is 150 mm square and 4 m high. How much will the post shorten when an axial load of 108 kN is applied?

Solution The modulus of elasticity E of this type of timber is known to be 8000 N/mm^2.

$A = 22\,500\text{ mm}^2 \quad l = 4000\text{ mm}$

Therefore,

$$\delta l = \frac{Wl}{AE} = \frac{108\,000 \times 4000}{22\,500 \times 800} = 2.4\text{ mm}$$

This problem demonstrates that if the modulus of elasticity of a material is known, the amount a given member will shorten (or lengthen) under a given load can be calculated.

Pull

Bar gripping device

Steel bar

Gauge length

Pull

Figure 6.7

Table 6.2

	Load kN	Elongation mm
1	0.0	0.00
2	3.8	0.04
3	7.6	0.08
4	11.4	0.12
5	15.2	0.16
6	19.0	0.20

BEHAVIOUR OF STEEL IN TENSION: YIELD POINT

A steel bar, 12 mm in diameter, was gripped in the jaws of a testing machine and subjected to a gradually increasing pull (Fig. 6.7). A gauge was attached to two points on the bar 250 mm apart (gauge length = 250 mm) and the following readings were recorded (Table 6.2). (The elongations given in the table have been slightly adjusted.)

The area A of a 12 mm diameter circle is 113.1 mm².

$$l = 250 \text{ mm}$$

From reading 2,

$$E = \frac{Wl}{A\,\delta l} = \frac{3800 \times 250}{113.1 \times 0.04} = 210\,000 \text{ N/mm}^2$$

From reading 3,

$$E = \frac{Wl}{A\,\delta l} = \frac{7600 \times 250}{113.1 \times 0.08} = 210\,000 \text{ N/mm}^2$$

and so on.

The constant value of E in this case is 210 000 N/mm².

Note that the modulus of elasticity of steel is much greater than that of timber. Steel, of course, is much more difficult to stretch than is timber.

Within the range of the above experiment, where the highest recorded load was 19 kN, the steel behaved as an elastic material. This means that, when the load is taken off, the bar will revert to its original length; on reloading, the bar will again elongate; and on unloading the elongation will disappear, and so on.

If the load is gradually increased beyond 19 kN, the bar will continue to stretch proportionately to the applied load until a loading of between 30 and 32 kN is reached. (These loads on a bar 12 mm in diameter are equivalent to stresses of 265–280 N/mm²). At about this point the steel reaches what is called its **elastic limit** (i.e. it ceases to behave as an elastic material) and begins to stretch a great amount compared with the previous small elongations. This stretching takes place without the application of any further load and the steel is said to have reached its **yield point.**

At this point the **plastic** or **ductile behaviour** of the steel begins. If the load is removed from the steel after it has been loaded beyond its yield point, the steel will not revert to its original length. The elongation which remains is called *permanent set.*

After a short time the steel recovers a little and ceases to stretch. Additional load can now be applied, but the steel has

Figure 6.8

Figure 6.9

been considerably weakened and stretches a great deal for each small increment of load. Finally, at about a stress of 450 N/mm² the bar breaks, but just before it breaks, it *waists* at the point of failure as indicated in Fig. 6.8. If the two fractured ends are placed together and the distance between the original gauge points is measured it will be found that the total elongation is 20 per cent or more of the original length.

Fig. 6.9 shows a graph which has been constructed from the results of a test on a 12 mm diameter bar, the gauge length being 250 mm.

Actually, the limit of proportionality (stress proportional to strain) is reached a little before the yield point, but for most practical purposes it is quite justifiable to consider the limit of proportionality, the elastic limit and the yield point to be identical. In addition, referring to Fig. 6.8, the bar reduces considerably in cross-section just before it fails, and some load can be taken off the bar so that the actual load which causes failure is less than the maximum recorded load. This is indicated on the graph by the dotted line. The ultimate or failing stress, i.e. the tensile strength of the steel, is calculated on the original cross-sectional area of the bar and not on the final reduced cross-section.

STRENGTH OF MATERIALS AND FACTOR OF SAFETY

In Examples 6.1 and 6.2 the stress in the steel was calculated as 90 N/mm² and the stress in the brickwork as 1.0 N/mm² (or 1.06 N/mm²) respectively. Two questions now arise:
a) Is it safe to allow these materials to be stressed to this extent, or
b) Can higher stresses be allowed so that smaller members can be used, thus economizing in material?

From the above discussion of the behaviour of steel in tension it can be deduced that the bar of 2000 mm² (Example 6.1) would require about 900 kN to cause it to fail in tension. The bar, however, is only supporting 180 kN, therefore it is amply strong and a smaller bar would suffice. The question now is: How small can the bar be? It would be, surely, unwise to make it of such dimensions that the steel in the bar would be stressed beyond its yield point because the steel would be in an unstable state. Various other factors also argue against using too high a stress for design purposes. The actual stress in the bar might be more (or even less) than the calculated value, due to assumptions made during the calculation stage. For example, the calculations may be based on the presence of a perfect hinge at a certain point in the structure. If the construction is such that there is a certain amount of fixity at the so-called hinge, the actual stresses in the members might be somewhat different from the calculated stresses.

Structural design is not an exact science, and calculated values of reactions, stresses, etc., whilst they may be mathematically correct for the theoretical structure (i.e. the model), may be only approximate as far as the actual behaviour of the structure is concerned.

For these and still other reasons it is necessary to make the design or working stress (or the allowable or permissible stress) less than the ultimate stress or (as intimated in the case of steel) the yield stress to allow for a safety margin against failure. This margin is provided by the introduction of the **factor of safety,** so that

$$\text{Design stress} = \frac{\text{Ultimate (or yield) stress}}{\text{Factor of safety}}$$

The value of the factor of safety varies between 1.4 and 2.4 at present and depends on many circumstances. It has been progressively reduced as the knowledge of structural behaviour of materials has increased and the quality of supervision of construction has improved.

The resulting design or permissible stresses for the various materials are specified in the appropriate British Standards and Codes of Practice. Their application to structural calculations is explained in later chapters.

In Chapter 1 an alternative method of ensuring the safety of a structure based on the ratio of loads was mentioned. It may be appropriate to explain here that, in the load factor method, the ultimate strength of the structure needs to be established. This means that the method requires the knowledge of plastic, as opposed to elastic, behaviour, which is outside the scope of this textbook, although reference to it is made in Chapter 10.

STRESSES IN COMPOSITE MEMBERS: MODULAR RATIO

The knowledge of the modulus of elasticity E is particularly useful in determining stresses in composite members. These are structural elements made up of two (or more) materials (e.g. steel and timber in flitch beams or steel and concrete in reinforced concrete beams), in which the materials are rigidly fixed together so that any changes in the length of the element are the same in each of the constituent materials.

Consider an element consisting of materials A and B. Then

$$E_A = \frac{\text{stress}_A}{\text{strain}_A} \quad \text{and} \quad E_B = \frac{\text{stress}_B}{\text{strain}_B}$$

but $\text{strain}_A = \text{strain}_B$, from the above definition of composite member, so taking the ratio of the moduli of elasticity

$$\frac{E_A}{E_B} = \frac{\text{stress}_A}{\text{stress}_B}$$

The ratio E_A/E_B, being the ratio of two moduli, is called the **modular ratio** and is denoted by the letter m.

$$\text{Modular ratio } m = \frac{E_A}{E_B}$$

Figure 6.10

Example 6.4

Two $150\,\text{mm} \times 75\,\text{mm} \times 4\,\text{m}$ long timber members are reinforced with a steel plate $150\,\text{mm} \times 6\,\text{mm} \times 4\,\text{m}$ long (Fig. 6.10), the three members being adequately bolted together.

The permissible stresses for the timber and the steel are $6\,\text{N/mm}^2$ and $130\,\text{N/mm}^2$, respectively, and E for timber is $8400\,\text{N/mm}^2$ and for steel is $210\,000\,\text{N/m}^2$.

Calculate the permissible tensile load for this composite member and the amount of elongation due to this load.

Solution

Area of timber $= 2 \times 150 \times 75 = 22\,500\,\text{mm}^2$

Area of steel $= 150 \times 6 = 900\,\text{mm}^2$

The stresses in the composite parts will be in the ratio

$$\frac{E_s}{E_t} = \frac{210\,000}{8400} = 25$$

so that if the stress in timber is $6\,\text{N/mm}^2$, the stress in steel would have to be $6 \times 25 = 150\,\text{N/mm}^2$ which exceeds the permissible stress for steel of $130\,\text{N/mm}^2$.

It follows, therefore, that the timber may not be fully stressed and the steel stress is the critical one.

Thus the stress in timber will be $130/25 = 5.2\,\text{N/mm}^2$. Hence

$$\text{Safe load for timber} = 5.2 \times 22\,500 = 117\,\text{kN}$$

$$\text{Safe load for steel} = 130 \times 900 \quad = 117\,\text{kN}$$

$$\overline{234\,\text{kN}}$$

and the elongation

$$\delta l = \frac{Wl}{AE} = \frac{117\,000 \times 4000}{900 \times 210\,000} = 2.5 \text{ mm or}$$

$$\delta l = \frac{117\,000 \times 4000}{22\,500 \times 8400} = 2.5 \text{ mm}$$

SUMMARY

Stress is obtained by dividing the applied load by the area of cross-section of the member, i.e.

$$\text{Tensile or compressive stress} = \frac{W}{A}$$

The design, permissible, working or allowable stress for a material depends on the nature of the material, the type of stress and the use of the material in the building. For example, whether it is used in a long column or in a short column.

Factor of Safety = failing or ultimate stress of the material divided by the design, permissible, working or allowable stress.

$$\text{Strain (tension or compression)} = \frac{\text{Change in length}}{\text{Original length}} = \frac{\delta l}{l}$$

In elastic materials obeying Hooke's Law stress is proportional to strain provided the elastic limit of the material is not exceeded.

Modulus of Elasticity or Young's modulus, E, is a measure of the resistance of an elastic material to being stretched or shortened. The greater the value of E, the more difficult it is to cause shortening or lengthening of the material.

$$E = \frac{\text{stress}}{\text{strain}} = \frac{Wl}{A\,\delta l}$$

and is measured in MN/m^2, N/mm^2, etc.

When two materials A and B are combined,

$$\frac{\text{Stress (A)}}{\text{Stress (B)}} = \frac{E(A)}{E(B)}$$

EXERCISE 6

1 A steel tie bar 100 mm × 10 mm in cross-section is transmitting a pull of 135 kN. Calculate the stress in the bar.

2 Calculate the safe tension load for a steel bar 75 mm × 6 mm in cross-section, the working stress being 155 N/mm^2.

3 A tie bar is 75 mm wide and it has to sustain a pull of 100 kN. Calculate the required thickness of the bar if the permissible stress is 150 N/mm^2.

Figure 6.11

75mm

x mm

75mm

Figure 6.12

150 mm

153 kN

4 A bar of steel circular in cross-section is 25 mm in diameter. It sustains a pull of 60 kN. Calculate the stress in the bar.

5 Calculate the safe load for a bar of steel 36 mm in diameter if the working stress is 155 N/mm².

6 A bar of steel, circular in cross-section, is required to transmit a pull of 40 kN. If the permissible stress is 150 N/mm² calculate the required diameter of the bar.

7 A timber tension member is 100 mm square in cross-section. Calculate the safe load for the timber if the permissible stress is 8 N/mm². Calculate the diameter of a steel bar which would be of equal strength to the timber member. Permissible stress for the steel is 150 N/mm².

8 A tie bar of the shape shown in Fig. 6.11 has a uniform thickness of 12 mm and has two holes 20 mm diameter each. Calculate the width x so that the bar is equally strong throughout its length. Calculate the safe pull for the bar if the permissible stress is 150 N/mm².

9 A tie bar of steel 150 mm wide is connected to a gusset plate by six rivets as shown in Fig. 6.12. The diameter of each rivet hole is 22 mm. Calculate the required thickness of the bar if the working stress is 155 N/mm².

10 Calculate the cross-sectional dimensions of a square brick pier to support an axial load of 360 kN, if the permissible stress for the brickwork is 1.7 N/mm².

11 A short specimen of deal timber 50 mm square in cross-section failed in a compression machine at a load of 70 kN. The permissible stress for such timber is 5.6 N/mm². Calculate the factor of safety.

12 A steel stanchion carrying a load of 877.5 kN is to be provided with a square steel base plate to spread the load on to a concrete foundation block. Calculate the minimum length of side (in mm) of the base plate if the stress on the concrete must not exceed 4.5 N/mm².

13 A steel column circular in cross-section is 150 mm in diameter and carries a load of 1.2 MN. Calculate a) the compressive stress in the column; b) the length of side of a square steel base plate to transmit the column load to a concrete foundation block; the permissible stress on the concrete is 4.5 N/mm²; c) assuming the concrete base to weigh 150 kN, calculate the plan dimensions of the concrete foundation so that the stress on the soil does not exceed 200 kN/m².

14 The following data were recorded during a tensile steel test:

Diameter of bar = 20 mm
Distance between gauge points = 200 mm
Elongation due to load of 50 kN = 0.18 mm
Load at yield point = 79 kN
Failing or ultimate load = 127 kN

Calculate, in N/mm^2, *a*) the stress at yield point; *b*) the ultimate stress; *c*) the modulus of elasticity of the steel.

15 A steel bar 100 mm × 12 mm in cross-section and 3 m long is subjected to an axial pull of 130 kN. How much will it increase in length if the modulus of elasticity of the steel is 210 000 N/mm^2?

16 A hollow steel tube 100 mm external diameter and 80 mm internal diameter and 3 m long is subjected to a tensile load of 400 kN. Calculate the stress in the material and the amount the tube stretches if Young's modulus is 200 000 N/mm^2.

17 During an experiment on a timber specimen 75 mm × 75 mm in cross-section, a shortening of 0.22 mm was recorded on a gauge length of 300 mm when a load of 36 kN was applied. Calculate the modulus of elasticity of the timber. Using this value of *E*, determine the amount of shortening of a timber post 150 mm square and 2.4 m high due to an axial load of 130 kN.

18 Assuming the permissible stress for a timber post 150 mm square and 2.7 m high is 6.5 N/mm^2, calculate the safe axial load for the post. How much will the post shorten under this load, assuming *E* to be 11 200 N/mm^2?

19 Three separate members of steel, copper and brass are of identical dimensions and are equally loaded. Young's moduli for the materials are: steel, 210 000 N/mm^2; copper, 100 000 N/mm^2; brass, 95 000 N/mm^2. If the steel member stretches 0.13 mm calculate the amount of elongation in the copper and brass members.

20 During a compression test, a block of concrete 100 mm square and 200 mm long (gauge length = 200 mm) shortened 0.2 mm when a load of 155 kN was applied. Calculate the stress and strain and Young's modulus for the concrete.

21 A tension member is made of timber and steel firmly fixed together side by side. The cross-sectional area of the steel is 1300 mm^2 and that of the timber is 4000 mm^2 and the length of the member is 3 m. If the maximum permissible stresses for the steel and timber when used separately are 140 N/mm^2 and 8 N/mm^2 respectively, calculate the safe load which the member can carry and the increase in length due to the load. Young's modulus for steel is 210 000 N/mm^2 and for timber is 8400 N/mm^2.

22 A structural member made of timber is 125 mm × 100 mm in cross-section. It is required to carry a tensile force of 300 kN and is to be strengthened by two steel plates 125 mm wide bolted to the 125 mm sides of the timber. Calculate the thickness of steel required if the permissible stresses for the steel and wood are 140 N/mm^2 and 7 N/mm^2 respectively. Assume that Young's modulus for the steel is 25 times that for the timber.

23 A timber post 150 mm square has two steel plates 150 mm × 6 mm bolted to it on opposite sides along the entire length of the post. Calculate the stresses in the timber and steel due to a vertical axial load of 350 kN. If the post is 3 m high calculate the amount of shortening under the load. *E* for steel is 210 000 N/mm^2 and *E* for timber is 8000 N/mm^2.

24 A metal bar consists of a flat strip of steel rigidly fixed alongside a flat strip of brass. The brass has a cross-sectional area of 900 mm² and the steel 300 mm². The compound bar was placed in a tensile testing machine and the extension measured by means of an extensometer fixed over a 250 mm gauge length. The extension was recorded as 0.13 mm. Calculate the load applied to the bar and the stress in each material.

Q2

205000

E for brass = 80 000 N/mm² E for steel = 210 000 N/mm²

7 Shear Force and Bending Moment

In this chapter the twin effects of the action of applied loads on beams are investigated in terms of the shear force and bending moment. The plotting of shear force (SF) diagrams and bending moment (BM) diagrams is demonstrated, and the significance of the position of the points of zero shear (maximum bending moment M_{max}) and zero bending (contraflexure) is explained.

When a beam is loaded, the applied loads have a tendency to cause failure of the beam, and whether or not the beam *actually* does fail depends obviously upon the extent or amount of loading and on the size and strength of the beam in question.

It is necessary to provide a beam that will safely carry the estimated loading with a reasonable factor of safety, and which will at the same time be light enough for economy and shallow enough to avoid unnecessary encroachment upon headroom.

In order to calculate the stresses that loading will induce into the fibres of a beam's cross-section, and to compare them with the known safe allowable stress for the material of which the beam is made, it is necessary to study the ways in which loading *punishes* a beam, and to assess the degree of *punishment.*

Loading tends to cause failure in two main ways:

1 By "shearing" the beam across its cross-section, as shown in Fig. 7.1.

2 By bending the beam to an excessive amount, as shown in Fig. 7.2.

These two tendencies to failure or collapse do occur simultaneously, but for a clearer understanding of each they will be examined separately.

Figure 7.1

Figure 7.2

Figure 7.3

Figure 7.4

SHEAR FORCE

Consider the portion of beam shown in Fig. 7.3(*a*). The loads tending to shear the beam are the two point loads acting downwards and the l.h. reaction acting upwards.

For simplicity of explanation the weight of the beam itself has been ignored.

Fig. 7.3(*b*) shows how the beam would tend to shear at point A. The only load to the left of A is the l.h. reaction (acting upwards) of 55 kN and there is therefore a resultant force of 55 kN tending to shear the portion of beam to the left of A upwards as shown. Note that the loads to the right of A are 70 kN downward and reaction 15 kN upward, so that there is also a resultant of 55 kN tending to shear the portion of beam to the right of A downward.

This 55 kN upward to the left, and 55 kN downward to the right, constitutes the shearing force at the point A.

It follows from the above that **shear force** may be defined as the algebraic sum of the loads to the left or to the right of a point.

This type of shearing force, where the resultant shear is upwards to the left and downwards to the right, may be called *positive shear*.

Consider now the point B as shown in Fig. 7.3(*c*). The resultant shear to the left is seen to be $55 - 40 = 15$ kN upward to the left, and $30 - 15 = 15$ kN downward to the right (again positive shear). At point C in Fig. 7.3(*d*), the shear to the left is $30 + 40 - 55 = 15$ kN down, and to the right equals the r.h. reaction of 15 kN upward. This type of shear, down to the left and up to the right, is called *negative shear*.

The above example of shear force ignored the weight of the beam and the only loading consisted of point loads—that is to say the loading was considered to be applied at a definite point along the span. Loads in actual fact are rarely applied in this fashion to structural members, but many loads applied to beams approximate to point loading and in design are considered as concentrated loads. The main beam shown in Fig. 7.4, for example, carries the reaction from one secondary beam and in addition has sitting on its top flange a short steel post; both these loads are applied to such a short length of the beam that they may be considered as point loads.

On the other hand, there are many cases where the loading is applied at a more or less uniform rate to the span of the beam. Such an example is found where a brick wall is carried on the top flange of a beam, or when a reinforced concrete slab sits upon a steel or reinforced concrete beam. Now, shear force has been described as the algebraic summation of loads to the left (or to the right) of a point. Therefore, the shear alters in such cases only where another point load

Figure 7.5

Figure 7.6

occurs. When the loading is uniformly distributed, however, as in Fig. 7.5, then the shear force will vary at a uniform rate also and a sudden jump in the value will only occur at the point of application of the one point load. The shear at point A (Fig. 7.5(a)) upward to the left will be

140 kN up − 30 kN down = +110 kN

Similarly the shear at point B (Fig. 7.5(b)) upward to the left is

140 kN up − 60 kN down = 80 kN (again positive)

The shear just to the left of C (Fig. 7.5(c)) is

140 − 90 = 50 kN positive

whilst just to the right of C when the 40 kN point load is included it is

140 − (90 + 40) = 10 kN positive

Thus it will be seen that uniform loads cause gradual and uniform change of shear, whilst point loads bring about a sudden change in the value of the shear force.

BENDING MOMENT

The degree of punishment in bending is measured as bending moment, and the amount of bending tendency is dependent upon the loads and upon the distance between them.

For example, the beam shown in Fig. 7.6 tends to split in bending under the 30 kN load because the l.h. reaction of 15 kN acting 1.5 m to the left has a clockwise bending tendency of $15 \times 1.5 = 22.5$ kN m, at the point where splitting of this type would most easily occur in this case.

If the beam had been of 10 m span instead of only 3 m the l.h. reaction (ignoring for a while the self-weight of the beam) would still be only 15 kN, but this time the bending tendency at the point of maximum stress would be $15 \times 5 = 75$ kN m.

Reverting to the beam shown in Fig. 7.6, which we have seen has a clockwise bending tendency to the left of point C of 22.5 kN m, there is also an anticlockwise bending tendency to the right of C of $15 \times 1.5 = 22.5$ kN m caused by the action of the r.h. reaction. At *any point* along the span of a simple beam of this type supported at its ends, the bending tendency to the left will always be clockwise, and that to the right anticlockwise but of the same amount. Thus at any point of such a beam the bending will be of the *sagging* type as shown in Fig. 7.2, and the fibres towards the lower face of the section will be subjected to tension.

Beam fibres tear at point of maximum bending moment

40 kNm

2m

C

20 kN

Figure 7.7

2m

1m

60kN 10kN

C

R_L =80 kN

4m

Figure 7.8

1m

60kN   10kN

80kN

3m

Figure 7.9

Resultant from shaded portion of uniform load 30kN/m

2m

30kN/m

C

R_L = 200 kN

4m

Figure 7.10

There are types of beams that bend in the opposite way. For example, a cantilever as shown in Fig. 7.7 has a *hogging* rather than a sagging tendency and obviously the moments in this type must be anticlockwise to the left and clockwise to the right.

To distinguish between these two types of bending it is normal to describe sagging as positive and hogging as negative.

From the foregoing example, which has been of the very simplest nature, it follows that **bending moment** may be described as the factor which measures the bending effect at any point of a beam's span due to a system of loading. The amount of bending moment is found by taking the moments acting to the left *or* to the right of the point concerned. The beam discussed and shown in Fig. 7.6 had only one load acting to the left of the point C at which the bending moment was calculated; where there are more than one load on the portion of beam concerned—as for example in the portion of beam shown in Fig. 7.8—then taking moments to the left of point C, the reaction has a clockwise bending effect and the two downward loads have anticlockwise bending tendency. The net bending moment at C will be the difference between these two types of moment, i.e. their algebraic sum. The bending moment at any point is, therefore, defined as: the algebraic sum of the moments caused by the forces acting to the left or to the right of that point.

In Fig. 7.8, the l.h. reaction exerts a clockwise moment of $80 \, \text{kN} \times 4 \, \text{m} = 320 \, \text{kN m}$ about point C and the downward loads exert moments in an anticlockwise direction equal to $60 \, \text{kN} \times 2 \, \text{m} = 120 \, \text{kN m}$ and $10 \, \text{kN} \times 1 \, \text{m} = 10 \, \text{kN m}$. Thus the bending moment at C is

$$+80 \times 4 - 60 \times 2 - 10 \times 1 = 320 - 120 - 10 = 190 \, \text{kN m}$$

The bending moment at the point of the 10 kN load in the above beam would be

$$80 \, \text{kN} \times 3 \, \text{m} - 60 \, \text{kN} \times 1 \, \text{m} = 240 - 60 = 180 \, \text{kN m}$$

as shown in Fig. 7.9. Note that in this case the 10 kN load is ignored, as, passing through the point concerned, it does not exert a moment about that point.

When, on the other hand, a uniformly distributed load is applied to the beam, the punishment caused is less than that exerted by a comparable point load. Therefore, it would be most uneconomical in design to treat such loads as being concentrated at their mid-point.

Where such uniform rate of loading occurs, as in the case of the portion of beam shown in Fig. 7.10 for example, only that portion of loading which lies to the left of the point C (shown shaded) need be considered when the moment at that point to the left is being calculated. The shaded portion of load is $30 \text{ kN/m} \times 4 \text{ m} = 120 \text{ kN}$, and its resultant lies halfway along its length (2 m from C). Therefore the moment about C of the uniform load is $120 \text{ kN} \times 2 \text{ m} = 240 \text{ kN}$ and the total bending moment at point C is

$$200 \times 4 - 120 \times 2 = 800 - 240 = 560 \text{ kN m}$$

SHEAR FORCE AND BENDING MOMENT DIAGRAMS

Shear force and bending moment have been described in general terms and it will have been seen that the values of both vary at different points along the span.

It is often desirable to show this variation by means of diagrams, which are really graphs, and these diagrams are called Shear Force Diagrams and Bending Moment Diagrams.

The following examples will serve to show how the diagrams of this type may be constructed for simple and more complex cases.

1 Fig. 7.11(a) shows a simply supported beam of span l carrying one point load of W kN at the centre of the span. Since the loading is symmetrical the reactions must be equal to each other, and each reaction will be $\frac{1}{2}W$ kN. Ignoring the self-weight of the beam, at any point C at x m from the left-hand reaction, the shear force to the left will be simply the value of the l.h. reaction, i.e. $\frac{1}{2}W$ kN upward to the left, and wherever point C lies between the l.h. end and the load, the shear will still be $+\frac{1}{2}W$ kN.

Similarly at any point to the right of the load as at point D, the shear to the left of D is W kN downward and $\frac{1}{2}W$ kN upward, which is equal to $\frac{1}{2}W$ kN down to the left or $\frac{1}{2}W$ kN up to the right (negative shear). These variations of the shear values are shown to scale on the **Shear Force Diagram** (SFD) (Fig. 7.11(b)). The vertical ordinate of the diagram at any point along the span shows the shear force at that point, and the diagram is drawn to two scales: a vertical scale of 1 mm = a suitable number of kilonewtons, and a horizontal scale of 1 mm = a suitable number of metres, being the same scale as that used in showing the span of the beam. Fig. 7.11(c) shows the variation of bending moment and constitutes a **Bending Moment Diagram** (BMD).

Obviously the moment at each end is zero, and at any point C at x m from the l.h. end the bending moment (summa-

Beam with central load point

Shear Force Diagram

Bending Moment Diagram

Figure 7.11

tion of moments to the left) is simply $\frac{1}{2}W \times x = \frac{1}{2}Wx$ kN m. This moment increases as x increases, and will reach a maximum amount of $\frac{1}{2}W \times \frac{1}{2}l = \frac{1}{4}Wl$ at the centre of span as shown.

Again the diagram is drawn horizontally to the same scale as the span of the beam, but this time vertically to a scale of 1 mm = a suitable number of kN m or N mm, and the vertical ordinate at any point along the span represents the bending moment at that point.

Comparing the two diagrams, it will be seen that the bending moment everywhere on the span is positive, and that the shear force changes its type from positive to negative at the point along the span where the bending moment reaches its maximum amount.

Fig. 7.12 shows a simply supported beam loaded with one single non-central point load, at distance a from one end and at b from the other.

Taking moments about the l.h. end,

r.h. reaction $\times l = W \times a$

r.h. reaction $= Wa/l$

Similarly, taking moments about the r.h. end,

l.h. reaction $\times l = W \times b$

l.h. reaction $= Wb/l$

At any point between the l.h. end and the point load,

Shear to the left $=$ l.h. reaction $= Wb/l$ (positive).

Similarly, at any point between the load and the r.h. end,

Shear to the right $=$ r.h. reaction of Wa/l (up to the right, thus negative shear)

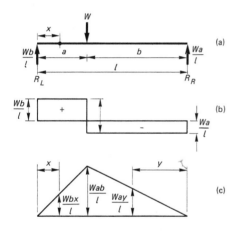

Figure 7.12

Thus the shear changes sign from positive to negative at the point load as in the previous example, and the SFD is as shown in Fig. 7.12(*b*).

As before, the bending moment is zero at the l.h. end, and at any point between that end and the load is equal to

$$R_L \times x = (Wb/l) \times x = Wbx/l$$

This reaches a maximum value of Wab/l at the point load (where $x = a$).

Also, at any point between the load and the r.h. end, at a distance of y from the r.h. reaction,

Bending moment $=$ r.h. reaction $\times y = Way/l$

and this also reaches a maximum at the point load (where $y = b$) of Wab/l.

The full bending moment diagram is thus a triangle with a

maximum vertical height of Wab/l at the load, and the bending moment at any point along the span may thus be scaled from the diagram to the same scale which was used in setting up the maximum ordinate of Wab/l.

2 The next case will deal with a simply supported beam of span l carrying a uniformly distributed load of intensity w kN/metre run (Fig. 7.13). The span consists of l m of load, and thus the total load is wl kN, and, as the beam is symmetrical, each reaction will be half of the total load

$$R_L = R_R = \tfrac{1}{2}wl \text{ kN}$$

The shear (up to the left) at a point just in the span and very very near to R_L is quite obviously simply the l.h. reaction of $\tfrac{1}{2}wl$.

When the point concerned is, say, 1 m from R_L however, the shear to the left (summation of loads to the left of the point) is then

$\tfrac{1}{2}wl$ upwards + 1 m of load w downwards

$= \tfrac{1}{2}wl - w$

Similarly when the point concerned is 2 m from R_L, the shear up to the left is

$\tfrac{1}{2}wl$ upwards − 2w (2 m of load) downwards

$= \tfrac{1}{2}wl - 2w$

Putting this in general terms, the shear at any point C on the span at distance x from R_L is

$\tfrac{1}{2}wl - x$ metres of uniform load $= \tfrac{1}{2}wl - wx$

This will give a positive result wherever x is less than $\tfrac{1}{2}l$, and a negative result where x exceeds $\tfrac{1}{2}l$, so the shear will "change sign" where $x = \tfrac{1}{2}l$ (at the point of mid-span), and the SFD will be as shown in Fig. 7.13(b).

Referring again to Fig. 7.13(a), the bending moment at the l.h. end is again zero, and at a point 1 m from R_L, the bending moment (summation of moment to the left) is simply the algebraic sum of

a) the clockwise moment of the l.h. reaction $(\tfrac{1}{2}wl \times 1)$;
b) the anticlockwise moment from 1 m of downward load $(w \times 1 \times \tfrac{1}{2} = \tfrac{1}{2}w)$.
Thus, the bending moment $= \tfrac{1}{2}wl - \tfrac{1}{2}w$.

If the bending moment is required at a point C, at x m from R_L, then the bending moment is the algebraic sum again of

a) the clockwise moment of the l.h. reaction $(\tfrac{1}{2}wl \times x)$;
b) the anticlockwise moment of x m of load $(wx) \times \tfrac{1}{2}x$
Thus, the bending moment $= \tfrac{1}{2}wlx - \tfrac{1}{2}wx^2$.

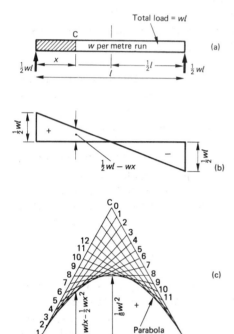

Figure 7.13

This bending moment will be positive for any value of x, and will reach a maximum value of

$$(\tfrac{1}{2}wl \times \tfrac{1}{2}l) - \tfrac{1}{2}w(\tfrac{1}{2}l)^2 = \tfrac{1}{4}wl^2 - \tfrac{1}{8}wl^2$$
$$= \tfrac{1}{8}wl^2$$

It should be most carefully noted that the maximum bending moment (M_{max}) is $\tfrac{1}{8}wl^2$, where w is the amount of uniform load per metre.

Sometimes it is more convenient to think in terms of the *total load* W (i.e. capital W), and in this case $W = wl$, and M_{max} in terms of the total load will then be

$$\tfrac{1}{8}wl \times l = \tfrac{1}{8}Wl$$

If the values of this bending moment at points along the span are plotted as a graph, the resulting BMD will be a parabola with a maximum ordinate of $\tfrac{1}{8}wl^2$ or $\tfrac{1}{8}Wl$ as shown in Fig. 7.13(c). Where the diagram has to be drawn, it will be necessary only to draw a parabola having a central height of $\tfrac{1}{8}Wl$, and any other ordinates at points away from the centre may be scaled or calculated as desired.

3 It is advisable, at this stage, to study carefully the Shear Force Diagrams already drawn and to note that these diagrams are drawn on a horizontal base. The upward loads (reactions) are projected upward on this base, and at the point loads, the SFD drops vertically by the amount of the point load.

Where a uniform load occurs, however, the SFD slopes down at a uniform rate, dropping for each metre of span an amount equal to the amount of uniform load per metre as shown in Fig. 7.14. It should be observed that in every case, the value of the bending moment has been a maximum at the point where the shear force changes its sign (generally referred to as the point of zero shear).

Bearing this in mind, it is common practice to draw the Shear Force Diagram simply to discover where the maximum bending moment occurs. This is quite unnecessary, as will be seen from consideration of the two cases shown in Figs. 7.15 and 7.16.

In Fig. 7.15, the l.h. reaction is 130 kN, and the shear force will change sign at the point of M_{max} where the downward loads starting from the l.h. end just equal 130 kN.

From A to C, the downward uniform load is 2 m at 15 kN/m = 30 kN. Then at the point load, there is a further drop of 70 kN, making a drop in the shear diagram from A to C of 100 kN.

Hence, if D is the point of M_{max} (zero shear) then the portion of load shown shaded must equal $130 - 100 = 30$ kN,

Figure 7.14

Figure 7.15

Figure 7.16

Figure 7.17

so that the downward loads up to D equal the l.h. reaction.

Thus, if the distance from C to D is x, then x m at 15 kN/m must equal 30 kN so $15x = 30$, i.e. $x = 2$ m and the maximum bending moment occurs at $2 + 2 = 4$ m from R_L.

Similarly in the case shown in Fig. 7.16 the load from the l.h. end up to and including the 70 kN point load is $2 \times 15 + 70 = 100$ kN.

Therefore, a further $145 - 100 = 45$ kN of load is required beyond C to D, the point of maximum bending moment. Therefore

$$15 \times x = 45 \text{ kN}$$

$$x = \frac{45}{15} = 3 \text{ m}$$

and M_{max} occurs at point D which is $2 + 3 = 5$ m from the l.h. end.

It will be seen that where the loading is uniform, the position of the maximum bending moment (M_{max}) occurs at a point on the span such that the sum of the downward loads from the l.h. end exactly equals the l.h. reaction.

Similarly, the downward loads taken from the r.h. end will equal the r.h. reaction at this point.

4 When the loading includes point loading, however, as it does in the example shown in Fig. 7.17, it may well be that the point at which the bending moment is a maximum coincides with the position of a point load—and in the beam illustrated in Fig. 7.17 the maximum value occurs at the 40 kN load.

Summing up the loads (starting from the l.h. end) it will be seen that just to the left of the 40 kN load they are $30 \times 1.7 + 10 = 61$ kN, that is just less than R_L.

Immediately to the right of the 40 kN load they add up to $30 \times 1.7 + 10 + 40 = 101$ kN which is more than R_L.

The rule for finding the position of the **maximum bending moment** may therefore be stated as follows: Add the downward loads together, starting from one reaction to the point where they equal (or suddenly become greater than) that reaction. This is the point of zero shear, and the point of maximum bending moment.

5 A beam which is supported at one end only by being firmly built into a wall or which is held horizontally at one end only by other means, is called a **cantilever.** Fig. 7.18 shows such a cantilever AB having a length of l m and loaded with one point load of W kN at the free end B.

Shear Force Diagram

Bending Moment Diagram

Figure 7.18

Shear Force Diagram

Bending Moment Diagram

Figure 7.19

The only downward load is W kN, so for equilibrium the l.h. upward reaction will also be W kN, but in addition, to prevent the beam from rotating, the wall or other form of restraint at the fixed end A must exert a moment of Wl in an anticlockwise direction.

The shear at any point C between A and B (sum of the loads to the left or to the right of C) will be W kN down to the right, or W kN up to the left. This is positive shear, and the value remains the same at all points along the span, so that the Shear Force Diagram is a rectangle as shown in Fig. 7.18(b).

The bending moment (taken as the sum of moments to the right) at any point C, distance x m from the free end B, will be the downward load of W multiplied by x m.

Thus the bending moment at any point is Wx kN m. This is a negative or hogging bending moment and its value varies directly as the value of x varies, so that the BMD will be a triangle as shown in Fig. 7.18(c).

The maximum value will occur at the fixed end when $x = l$, and its amount is Wl kN m.

Note: The previous rule (page 106) for finding the position of M_{max} will not apply to cantilevers. In the case of cantilevers M_{max} will always occur at the fixed end.

6 Fig. 7.19 shows a cantilever with a uniformly distributed load of w kN/m. The total load will be $w \times l = wl$ kN and the upward reaction at A will also be wl kN.

The moment of the downward load wl about the support A is

$$wl \times \tfrac{1}{2}l = \tfrac{1}{2}wl^2 \text{ kN m}$$

Thus the wall (or other form of restraint) at A must exert an anticlockwise moment of $\tfrac{1}{2}wl^2$ on the beam to prevent its rotating under the couple formed by the upward reaction and the resultant of the downward load.

The shear at any point C at distance x from the free end will be the portion of load to the right of C which is x m at w kN/m $= wx$ kN (positive shear). This varies at the same rate as x varies, and so the shear force diagram will be a triangle having a maximum value at the reaction (where $x = l$) of wl kN, as shown in Fig. 7.19(b).

Similarly the bending moment at C will be the moment (clockwise) of the portion of load to the right of C which is $(wx) \times \tfrac{1}{2}x = \tfrac{1}{2}wx^2$ kN m (negative or sagging moment).

This again reaches a maximum value of $\tfrac{1}{2}wl^2$ kN m at the fixed end A where $x = l$ and the rate of increase will be found to form a parabola as shown in the Bending Moment Diagram in Fig. 7.19(c).

Note: If the total load W is used instead of the value of the load per metre run (w), then $W = wl$, and the maximum bending moment will be $\frac{1}{2}Wl$ kN m.

Example 7.1

Draw the SF and BM Diagrams for the cantilever shown in Fig. 7.20 indicating all important values.

Solution

Reaction There is only one reaction, R_L at A

$$R_L = 80 + 40 + 30 + 20 - 76 = 94 \text{ kN}$$

Shear force

At any point between A and C, the shear downward to the right is the sum of the three downward loads which is 60 kN positive.

Similarly between C and D the shear is the sum of the two loads to the right which is 30 kN positive. Between D and the free end B the shear is the single load of 10 kN positive.

This is shown in the Shear Force Diagram constructed in Fig. 7.20(b).

Bending moment

The Bending Moment Diagram is most easily drawn by treating the three loads separately: the 10 kN load produces a bending moment of $10 \times 4 = 40$ kN m at A and the BMD for this load alone is the triangle 1 shown in Fig. 7.20(c).

The 20 kN load causes a bending moment at A of $20 \times 3 = 60$ kN m and the BMD for this load alone is the triangle 2.

Finally, the 30 kN load causes a $30 \times 1 = 30$ kN m bending moment at A and the BMD for this load is the triangle 3.

As in fact all three loads are on the beam at the same time, then the final BMD is the sum of the three triangles 1, 2 and 3 as shown in Fig. 7.20(c).

Figure 7.20

Example 7.2

Draw the SF and BM Diagrams for the beam loaded as shown in Fig. 7.21(a) indicating all important values.

Solution

Reactions

Taking moments about A

$$R_R \times 4 = 20 \times 4 \times \tfrac{1}{4} 4 + 40 \times 0.8 + 30 \times 1.6 + 20 \times 3.2$$
$$= 160 + 32 + 48 + 64 = 304$$

Figure 7.21

Final Shear Force Diagram

Figure 7.22

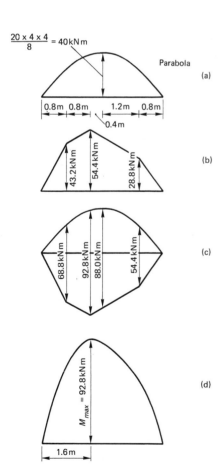

Figure 7.23

$$R_R = 76 \text{ kN}$$

and $\quad R_L = 80 + 40 + 30 + 20 - 76 = 94 \text{ kN}$

Shear force

Note: It should be obvious by now that the SF diagrams represent all point loads as vertical lines—upward in the case of reactions, and downward in the case of downward loads.

Also downward uniform loads are shown as sloping lines (gradual change in the value of the shear force) and therefore the SF Diagram may be plotted quickly from a horizontal base by merely plotting the loads in this way as they occur.

The horizontal base is shown as line A1–B1 in Fig. 7.21(b). The procedure is as follows:

1 Draw up from A1 the vertical reaction R_L of 94 kN (A1–F).

2 Draw the gradual change of (20×0.8) kN between A and C as the sloping line F–G.

3 Draw G–H of 40 kN vertically downward to represent the sudden change of shear at the 40 kN load at point C.

4 Draw the gradual change of 16 kN between C and D as the sloping line H–J.

5 Draw J–K of 30 kN vertically downward to represent the sudden change of shear at the 30 kN load at point D.

6 Draw the gradual change of (20×1.6) kN between D and E as the sloping line K–L.

7 Draw L–M of 20 kN vertically downward to represent the sudden change of shear at the 20 kN point load at E.

8 Draw the gradual change of 16 kN between E and B as the sloping line M–P.

9 Finally draw the vertical right hand reaction of 76 kN upward from P to join the horizontal base at B1.

Fig. 7.21(b) shows the construction of this Shear Force Diagram, and the final diagram with its important values is shown in Fig. 7.22.

Bending moment

The bending moment at any point or the maximum bending moment may be calculated in the usual way by taking moments to the left or to the right of that point. But if the final BMD is required (as it is in this example) then it may be drawn by constructing a) the BMD for the uniform load (Fig. 7.23(a)) and b) the BMD for the point loads alone (Fig. 7.23(b)) and adding the two diagrams together as shown in Fig. 7.23(c), resulting finally in the combined BMD given in Fig. 7.23(d).

The value of the bending moment at any point may also be obtained by computing the area of the SF Diagram to the left or to the right of that point.

Consider the bending moment at the centre of the span. The positive shear force area is

$$\tfrac{1}{2}(94+78)\times0.8+\tfrac{1}{2}(38+22)\times0.8$$
$$=68.8+24.0=92.8 \text{ kN m}$$

To obtain the negative shear force area, it is necessary to determine the value of the shear force at the centre of the span which is 0.4 m to the right of the 30 kN point load.

So, the shear force at the centre of span is

$$8+20\times0.4=16 \text{ kN}$$

Hence the negative shear force area to the left of the centre of the span is

$$\tfrac{1}{2}(8+16)\times0.4=4.8 \text{ kN m}$$

Thus the bending moment at the centre of the span is

$$92.8-4.8=88.0 \text{ kN m}$$

Example 7.3

Draw the Shear Force Diagram for the beam shown in Fig. 7.24, and calculate the value of M_{max}. Determine also the bending moment at a point C at 2 m from the l.h. reaction.

Solution Taking moments about A to determine reactions:

$$R_R \times 9 = 30\times9\times(\tfrac{1}{2}\times9)+60\times6\times(\tfrac{1}{2}\times6)$$
$$= 1215+1080 = 2295 \text{ kN}$$
$$R_R = 255 \text{ kN}$$

and

$$R_L = 60\times6+30\times9-255 = 375 \text{ kN}$$

M_{max} will occur at x m from A where

$$x \text{ m}\times(60\times30) \text{ kN/m} = R_L = 375 \text{ kN}$$
$$x = \frac{375}{60+30} = 4.167 \text{ m}$$

The value of M_{max} will, therefore, be

$$M_{max} = 375\times4.167-(60+30)\times4.167\times(\tfrac{1}{2}\times4.167)$$
$$= 1562.50-781.25 = 781.25 \text{ kN m}$$

or by computing the areas of the SF Diagram:

$$M_{max} = 375\times(\tfrac{1}{2}\times4.167) = 781.25 \text{ kN m (as before)}$$

Figure 7.24

Bending moment at C, 2 m from R_L, is

$$375 \times 2 - (30 + 60) \times 2 \times (\tfrac{1}{2} \times 2) = 750 - 180 = 570 \text{ kN m}$$

If a Bending Moment Diagram is required, bending moment values may be calculated at several points of the span and these values drawn to scale vertically from a horizontal line representing the span of the beam. The BMD is obtained by joining the tops of these lines.

Example 7.4

(a)

Shear Force Diagram

(b)

Figure 7.25

Draw the Shear Force Diagram for the loaded beam shown in Fig. 7.25, and determine the position and amount of the maximum bending moment (M_{max}).

Solution Taking moments about A to determine reactions,

$$R_R \times 12 = 36 \times 3 \times 1.5 + 18 \times 12 \times 6 + 12 \times 3 \times 10.5$$
$$= 162 + 1296 + 378 = 1836$$
$$R_R = 153 \text{ kN}$$

and

$$R_L = 36 \times 3 + 18 \times 12 + 12 \times 3 - 153$$
$$= 108 + 216 + 36 - 153$$
$$= 207 \text{ kN}$$

Position of M_{max}

At point C, 3 m from l.h. end, downward load from l.h. end is

$$(36 + 18) \times 3 = 162 \text{ kN}$$

M_{max} will occur where this has increased to 207 kN. Thus the M_{max} will occur at point D where the amount of load between C and D (shown shaded) is

$$207 - 162 = 45 \text{ kN}$$

Thus

Distance $y \times 18 \text{ kN/m} = 45 \text{ kN}$

$y = 45/18 = 2.5 \text{ m}$

Therefore M_{max} occurs at D at $(3.0 + 2.5) = 5.5$ m from the l.h. end:

$$M_{max} = 207 \times 5.5 - 162 \times (2.5 + \tfrac{1}{3}3.0) - 18 \times 2.5 \times \tfrac{1}{2}2.5$$
$$= 1138.5 - 648 - 56.25$$
$$= 434.25 \text{ kN m}$$

Check by shear force areas:

$$M_{max} = \tfrac{1}{2}(207 + 45) \times 3 + 45 \times \tfrac{1}{2}2.5$$
$$= 378 + 56.25$$
$$= 434.25 \text{ kN m}$$

Figure 7.26

Example 7.5

Draw the Shear Force and Bending Moment Diagrams for the beam shown in Fig. 7.26, and determine the position and amount of the maximum bending moment.

Solution Taking moments about A

$$R_R \times 1.8 = 45 \times 0.6 + 54 \times 1.2 + 18 \times 2.4$$
$$= 27.0 + 64.8 + 43.2$$
$$= 135$$
$$R_R = 75 \text{ kN}$$

and $\quad R_L = 45 + 54 + 18 - 75 = 42 \text{ kN}$

The shear force diagram is as shown in Fig. 7.25(*b*). The values of bending moments at

45 kN load $= 42 \times 0.6 = 25.2$ kN m		$(+)$
54 kN load $= 42 \times 1.2 - 45 \times 0.6 = 23.4$ kN m		$(+)$
r.h. support $= 42 \times 1.8 - 45 \times 1.2 - 59 \times 0.6$		
$= 10.8$ kN m		$(-)$

The bending moment diagram is as shown in Fig. 7.26(*c*).

Note that, due to the overhanging of the r.h. end, part of the beam is subjected to negative bending moment.

There are, therefore, two points of zero shear: one at C which marks the point of maximum positive moment, and one at E where the maximum negative moment occurs. The absolute maximum bending moment is thus 25.2 kN m at C, and the shape of the deflected beam is as shown in Fig. 7.26(*d*).

The point at which the negative bending moment changes to positive (and vice versa) is called the **point of contraflexure**. The value of the bending moment at that point is zero.

Example 7.6

Draw the Shear Force and Bending Moment Diagrams for the beam shown in Fig. 7.27 and determine the position and amount of the maximum bending moment.

Figure 7.27

Figure 7.28

Solution Since the loading and the beam are symmetrical, the reactions are each equal to $\frac{1}{2}(20+70+20) = 55$ kN.

Bending moment at D $= 55 \times 1.3 - 20 \times 2.3$

$$= 71.5 - 46.0$$

$$= 25.5 \text{ kN m} \quad (+)$$

Bending moment at C $= -20 + 1.0$

$$= -20.0 \text{ kN m} \quad (-)$$

Example 7.7

A beam 4 m long carrying a uniformly distributed load of 60 kN/m cantilevers over both supports as shown in Fig. 7.28(a). Sketch the SF and BM Diagrams and determine the position of the point of contraflexure.

Solution The shear force and bending moment diagrams are given in Fig. 7.28(b) and (c). Check that the reactions are 96 kN and 144 kN respectively. The maximum positive bending moment occurs at 1.6 m from the left end of the beam and is

$$M_{max} = \frac{1}{2}(66 \times 1.1 - 30 \times 0.5)$$

$$= \frac{1}{2}(72.6 - 15.0)$$

$$= 28.8 \text{ kN m}$$

The maximum negative bending moment occurs over the right-hand support and is $60 \times 0.5 = 30$ kN m.

By calculating several other bending moment values the diagram can be constructed as shown in Fig. 7.28(c).

Bending moment at point of contraflexure is

$$96(x - 0.5) - 60 \times x \times \tfrac{1}{2}x = 0$$

$$5x^2 - 16x + 8 = 0$$

$$x = \frac{16 \pm \sqrt{(256 - 160)}}{10}$$

$$x_1 = 0.62 \text{ m} \qquad x_2 = 2.58 \text{ m}$$

SUMMARY

Shear force at a point in the span of a beam is defined as the algebraic sum of the loads to the left or to the right of the point.

Bending moment at a point in the span of the beam is defined as the algebraic sum of all the moments caused by the loads acting to the left or to the right of that point.

The **maximum bending moment** (M_{max}) is the algebraic sum of the moments of the loads and reaction acting to *one side* of the point of maximum bending moment.

To determine the position of the maximum bending moment in a beam simply supported at its two ends, add together the downward loads starting from one reaction to the point where they equal (or suddenly become greater than) that reaction. This is the point of zero shear and the point of maximum bending moment.

The maximum bending moment in cantilevers will always occur at the fixed end and equals the sum of the moments of all the loads about the fixed end.

The **point of contraflexure** occurs where the bending moments change sign and their value is zero.

EXERCISE 7

1–14 Calculate the reactions, determine the position and amount of the maximum bending moment for the beams shown in Figs. 7.29 to 7.42.

15 Fig. 7.43 shows the shear force diagram for a loaded beam. Sketch the beam, showing the loading conditions, and calculate the maximum positive and negative bending moments.

16 A steel beam spans 4.8 m and carries on its whole length a brick wall 220 mm thick and 2.7 m high. The brickwork weighs 20 kN/m³ and the self-weight of beam and casing is estimated as being 5.7 kN. What is the maximum bending moment on the beam?

Figure 7.29

Figure 7.30

Figure 7.31

Figure 7.32

Figure 7.33

Figure 7.34

Figure 7.35

Figure 7.36

Figure 7.37

Figure 7.38

Figure 7.39

Figure 7.40

Figure 7.41

Figure 7.43

Figure 7.42

Figure 7.44

Figure 7.45

Figure 7.46

Figure 7.47

Figure 7.48

Figure 7.49

Figure 7.50

17 A steel beam is as shown in Fig. 7.44. The portion between supports A and B carries a uniform load of 30 kN/m and there are point loads at the free ends as shown. What is the length l in metres between A and B if the bending moment at a point C midway between these supports is just zero?

18 For the beam loaded as shown in Fig. 7.45: a) calculate the reactions; b) determine the position and amount of the maximum positive and maximum negative bending moments.

19 Referring to Fig. 7.46, calculate a) end reactions; b) position and amount of M_{max}.

20 Calculate the maximum bending moment in Fig. 7.47.

21 Determine the position and amount of the maximum shear force and bending moment in Fig. 7.48.

22 Referring to Fig. 7.49, a) calculate the end reactions; b) determine the position and amount of the maximum bending moment, M_{max}.

23 For the cantilever loaded as shown in Fig. 7.50, calculate a) the bending moment at the 40 kN load; b) the bending moment at the point D; c) the M_{max}.

24 Calculate the maximum bending moment for the cantilever shown in Fig. 7.51.

25 A steel beam simply supported at its ends carries a load of varying intensity as shown (Fig. 7.52). Determine a) the end reactions R_L and R_R; b) the M_{max}; c) the bending moment at point C.

26 A steel post cantilevers vertically carrying two point loads as shown in Fig. 7.53. Calculate the bending moments at a) the 35 kN load; b) the base; c) 1 m from the base.

Figure 7.51

Figure 7.52

Figure 7.53

Figure 7.54

Figure 7.55

Figure 7.56

27 A steel beam loaded as shown in Fig. 7.54 has a maximum bending moment (occurring at the point load) of 135 kN m. What is the value W kN of the point load?

28 In the beam shown in Fig. 7.55, the maximum bending moment occurs at the supports R_L and R_R and the bending moment at the central 20 kN load is zero. What is the length in metres of span l?

29 In the beam shown in Fig. 7.56, the maximum negative bending moment is twice the amount of the maximum positive bending moment. What is the value in kN of the central load W?

30 Derive an expression for the beam as shown in Fig. 7.57 for a) the maximum bending moment; b) the bending moment at C.

Figure 7.57

8 Properties of Sections

This chapter considers the effect of the shape or profile of a beam's section on the beam's resistance to the punishing forces and moments induced by the loading. The beam itself must, of course, be made just strong enough to withstand these punishing effects with a reasonable factor of safety, and the strength of the beam or its degree of resistance to bending moment and shear force is built up in terms of

a) the shape and size of the beam's section;

b) the strength of the particular material of which the beam is made.

The final degree of "defence" or resistance will be measured in units which take into account both of these two factors, but for a clearer understanding they will be treated separately to begin with. The present chapter deals with *a*) only, and will show how the shape or profile of a beam's section affects its strength.

The properties which various sections have by virtue of their shape alone are

1 Cross-sectional area.

2 Position of centre of gravity or area (centroid).

3 Moment of inertia or the second moment of area.

4 Section modulus.

5 Radius of gyration.

The properties *section modulus* and *radius of gyration* will be considered later in Chapters 9 and 12 respectively.

The cross-sectional area should need no description and may be calculated with ease for most structural members.

THE CENTRE OF GRAVITY OR CENTROID

The *centre of gravity* of a body is a point in or near the body

Figure 8.1

Figure 8.2

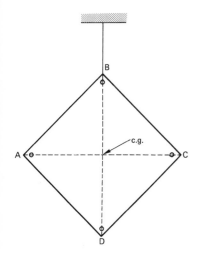

Figure 8.3

through which the resultant attraction of the earth, i.e. the weight of the body, acts for all positions of the body.

It should be noted, however, that the section of a beam is a plane figure without weight, and therefore the term *centre of area* or *centroid* is more appropriate and is frequently used in this case. The determination of the position of the centre of gravity of a body or centroid of a section is equivalent to determining the resultant of a number of like, parallel forces.

Fig. 8.1 shows a thin sheet of tinplate with several small holes drilled in it. To one of these holes (hole A) is attached a string. It should be obvious that the position shown for the sheet in Fig. 8.1 is an impossible one. The sheet will swing round in a clockwise direction and come to rest as indicated in Fig. 8.2. Each particle of the sheet in Fig. 8.1 is attracted vertically downwards by the force of gravity and the parallel lines indicate the direction of the gravity forces. The resultant of these parallel forces is the total weight of the sheet. The sheet comes to rest in such a position that the line of the resultant weight and the line of the vertical reaction in the string form one continuous line. When the string is attached to point B the sheet will hang as in Fig. 8.3 and the intersection of the two lines AC and BD is called the centre of gravity of the body. The actual position of the centre of gravity is in the middle of the thickness of the metal immediately behind the intersection of the two lines AC and BD. If the thickness be infinitely reduced, as it is in the case of a beam section, the position of the centre of gravity will coincide with that of the centroid (centre of area).

The position of the centre of gravity of sheets of metal of various shapes can be obtained by the method shown in Fig. 8.4. The sheet is suspended from a pin and from the same pin is suspended a plumb line. The line of the string forming the plumb line can be marked on the sheet behind it with pencil or chalk. The sheet can now be suspended in a different position and another line marked. The intersection of these two lines gives the position of the centre of gravity of the sheet.

The position of the centre of gravity of certain simple shapes is obvious by inspection (Fig. 8.5), e.g. the centre of gravity of a line or uniform rod is halfway along its length and the centre of gravity of a circle is at its centre.

If it were required to balance any of these shapes in a horizontal position by placing a pencil underneath, the point of application of the pencil would be the centre of gravity of the sheet, and the reaction of the pencil would be equal to the weight of the sheet. Each of the shapes shown in Fig. 8.5 has at least two axes of symmetry and whenever a body has two axes of symmetry the centre of gravity is at the intersection of the axes.

Figure 8.4

Front View   Side View

Figure 8.5

Figure 8.6

The centre of gravity of a body need not necessarily be in the material of the body (see Fig. 8.6). Note that it is impossible to balance this shape on one point so that the sheet lies in a horizontal plane.

To determine the position of the centre of gravity of a compound body or the centroid of a compound section which can be divided into several parts, such that the centres of gravity and weights of the individual parts are known, the following method applies:

1 Divide the body into its several parts.

2 Determine the area (or volume or weight) of each part.

3 Assume the area (or volume or weight) of each part to act at its centre of gravity.

4 Take moments about a convenient point or axis to determine the centre of gravity of the whole body. The method is identical with that of determining the resultant of a number of forces and is explained in the following example.

Figure 8.7

Example 8.1

A thin uniform sheet of material weighs *w* newtons for each mm² of its surface (Fig. 8.7). Determine the position of its centre of gravity.

Solution Since the figure is symmetrical about line AB, its centre of gravity must lie on this line. The figure can be divided into three rectangles.

Area (1) = $60 \times 150 = 9000$ mm²   Weight = $9w$ kN

Area (2) = $200 \times 20 = 4000$ mm²   Weight = $4w$ kN

Area (3) = $20 \times 100 = 2000$ mm²   Weight = $2w$ kN

$$\begin{array}{ll} \dfrac{\text{Total}}{\text{area}} = 15\,000 \text{ mm}^2 & \dfrac{\text{Total}}{\text{weight}} = 15w \text{ kN} \end{array}$$

Let $\bar{x}$ be the distance of the centre of gravity of the whole figure from O. Take moments about this point.

$$15w\bar{x} = 9w \times 30 + 4w \times 160 + 2w \times 270$$
$$= 270w + 640w + 540w$$
$$= 1450w$$
$$\bar{x} = 1450w/15w = 97 \text{ mm}$$

The centre of gravity is therefore on the line AB at 97 mm from O at the extreme left edge of the figure.

Note that the centre of gravity has been determined without knowing the actual weight of the material, and when a body is of uniform density throughout, its weight may be ignored and moments of areas or moments of volumes can be taken. The position of the centroid of a section is of great importance in beam design for, as will be seen later, the portion of section above this centroid performs a different function to that below the centroid.

Example 8.2

Determine the position of the centre of gravity of the body shown in Fig. 8.8. The body has a uniform thickness of 100 mm and weighs 10 N/m³.

Solution Since the body is homogeneous (i.e. of uniform density) and is of uniform thickness throughout, its weight and thickness may be ignored and moments taken of areas.

Area (1) = $\frac{1}{2}(3.0 \times 1.5) = 2.25$ m²

Area (2) = 4.0×1.5 $= 6.00$ m²

Area (3) = 1.0×3.0 $= 3.00$ m²

Total area = 11.25 m²

Figure 8.8

Figure 8.9

In centre of gravity problems it is usually convenient to choose two axes A–A and B–B at one extreme edge of the figure.

Let $\bar{x}$ be the horizontal distance of the centre of gravity of the whole figure from A–A and $\bar{y}$ be the vertical distance of the centre of gravity from axis B–B. The positions of the centre of gravity of the three separate parts of the figure are shown in Fig. 8.8.

Taking moments about axis A–A

$$11.25\bar{x} = 3.00 \times 1.5 + 6.00 \times 2.25 + 2.25 \times 2.5$$
$$= 4.5 + 13.5 + 5.6 = 23.6$$
$$\bar{x} = 23.6/11.25 = 2.1 \text{ m}$$

Taking moments about axis B–B

$$11.25\bar{y} = 3.00 \times 0.5 + 6.00 \times 3.0 + 2.25 \times 6.0$$
$$= 1.5 + 18.0 + 13.5 = 33.0$$
$$\bar{y} = 33.0/11.25 = 2.9 \text{ m}$$

The centre of gravity is in the position indicated by Fig. 8.9 and is within the thickness of the body halfway between the front and back faces

Use of the Link Polygon

Students who prefer a graphical method of solution to a calculation method can use the principle of the link polygon to obtain the position of the centre of gravity. The graphical solution of the previous problem is shown in Fig. 8.10, where the body is drawn to scale.

Example 8.3

A compound girder is built up of a 254×146UB37 steel beam with one 250 mm $\times$ 24 mm plate on the top flange only. The area of the steel beam alone is 4750 mm². Determine the position of the centroid of the compound girder (Fig. 8.11).

Solution

1 The compound section is symmetrical about the vertical axis Y–Y, so the centroid must lie on this line.

2 The addition of a plate to the top flange has the effect of moving the centroid (the X–X axis) towards that plate, and taking moments about line B–B at the lower edge of the section,

Total area of section $\times \bar{y} =$ (area of UB $\times$ 128 mm)

$$+ \text{(area of plate} \times 268 \text{ mm)}$$

Figure 8.10

Figure 8.11

$$(4750 + 6000)\bar{y} = 4750 \times 128 + 6000 \times 268$$
$$10\,750\bar{y} = 608\,000 + 1\,608\,000$$
$$= 2\,216\,000$$
$$\bar{y} = 2\,216\,000/10\,750 = 206 \text{ mm}$$

MOMENT OF INERTIA

The moment of inertia or, more appropriately in the case of a beam section (which, as stated earlier, is a plane figure without weight), the **second moment of area** of a shape, is a property which measures the *efficiency* of that shape in its resistance to bending.

Other factors besides shape enter into the building up of a beam's resistance to bending moment; the material of which a beam is made has a very obvious effect on its strength, but this is allowed for in other ways. The moment of inertia takes no regard of the strength of the material; it measures only the manner in which the geometric properties or shape of a section affect its value as a beam.

Figure 8.12

Figure 8.13

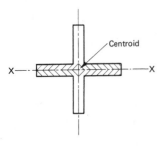

Figure 8.14

A beam such as the one shown in Fig. 8.12 may, in theory, be used with the web vertical or with the web horizontal, and it will offer much more resistance to bending when fixed as in (a) than when as in (b). This is because the moment of inertia about the X–X axis of bending is larger than the moment of inertia about the Y–Y axis.

These axes of symmetry X–X and Y–Y are termed the **principal axes** and they are in fact axes which intersect at the centroid of the section concerned.

It will normally be necessary to calculate the moments of inertia about both these principal axes, and these are usually described as I_{xx} and I_{yy}, the moments of inertia about the X–X and Y–Y axes respectively.

The manner in which the "build up" of a shape affects its strength against bending must—to be understood completely—involve the use of calculus, but it may still be made reasonably clear from fairly simple considerations.

The term moment of inertia is itself responsible for a certain amount of confusion. Inertia suggests laziness in some ways, whereas in fact the true meaning of inertia may be described as the resistance which a body makes to any forces or moments which attempt to change its shape (or its motion in the case of a moving body). A beam tends to change its shape when loaded, and inertia is the internal resistance with which the beam opposes this change of shape. The moment of inertia is a measure of the resistance which the section can supply in terms of its shape alone.

Good and Bad Beam Shapes

In order to make the best use of the beam material, the "shape" of the section has to be chosen with care. Certain shapes are better able to resist bending than others, and in general it may be stated that a shape will be more efficient when the greater part of its area is as far as possible away from its centroid.

A Universal Beam (UB) section (Fig. 8.13), for example has its flanges (which comprise the greater part of its area) well away from the centroid and the axis of bending X–X, and is in consequence an excellent shape for resisting bending.

Fig. 8.14 shows a shape built up by riveting together four angles. The bulk of its area (shown shaded) is situated *near to* the centroid, and the section is not a good one from the point of view of resistance to bending about the X–X axis.

If the same four angles are arranged as shown in Fig. 8.15, however, then the bulk of the area has been moved away from the axis of bending X–X, and the section—which now resembles roughly a UB form—is a good one, with a high resistance to bending.

Figure 8.15

Figure 8.16

Figure 8.17

Figure 8.18

When again the same four angles are used with a plate placed between them as in Fig. 8.16, the flanges are moved even further from the neutral axis, and the efficiency of the shape is much greater. Large deep girders are built up in this way when very heavy loads have to be supported.

When a shape such as that shown in Fig. 8.15 is being designed, it will be found that doubling the area of the *flange* will approximately double the flange's efficiency, but doubling the depth *d* as in Fig. 8.16 will increase the efficiency by 4, i.e. 2^2. The efficiency of the flange, then, varies directly as its area and as the square of its distance from the centroid.

Calculation of the Moment of Inertia

In order to measure the increase in efficiency of a section as its area and depth increase, the moment of inertia (second moment of area) is calculated, and this property does in fact measure both the direct increase in area and the square of the increase in depth.

To determine the moment of inertia of the rectangle shown in Fig. 8.17, it will be necessary to divide the shape into a number of strips of equal area as shown.

The area of each strip will be multiplied by the square of the distance of its centroid from the centroid of the whole section. This in fact amounts to assessing its second moment or, in other words, forming a factor which measures its area directly and the square of its distance from the axis of bending.

The sum of all such products $a \times y^2$ will be the moment of inertia (second moment of area) of the whole shape about the axis X–X.

Area $a \times y^2$ is the moment of inertia (second moment of area) of the shaded strip about axis X–X. The sum of all such products equals total I_{xx} of section.

Assume a rectangle to be as shown in Fig. 8.18, 180 mm wide and 320 mm in depth. Fig. 8.18 shows the section divided into 8 strips each 40 mm deep, each strip having an area of 7200 mm^2.

The sum of all the products

$$\sum ay^2 = \sum 2 \times a(y_1^2 + y_2^2 + y_3^2 + \ldots)$$
$$= 2 \times 7200(20^2 + 60^2 + 100^2 + 140^2)$$
$$= 14\,440 \times 33\,600$$
$$= 483.84 \times 10^6 \text{ mm}^4 \text{ units}$$

This is the approximate value of the I_{xx}, or moment of inertia about the X–X axis, but the exact value of this factor will of course depend upon the number of strips into which the shape is divided. It can be shown that if b is the width of the rectangle and d is the depth, then the exact value of I_{xx} is given by

$$I_{xx} = \tfrac{1}{12}bd^3$$

In the case above this is

$$\tfrac{1}{12} \times 180 \times 320^3 \quad \text{or} \quad 491.52 \times 10^6 \ \text{mm}^4 \ \text{units}$$

The units of the answer are important, and it should be appreciated that, as an area (mm²) has been multiplied by a distance squared (y^2), the answer is the second moment of area or more simply the moment of inertia of the shape, and is measured in mm⁴ units.

For those students who have an elementary knowledge of calculus, the foregoing approximate derivation of $\tfrac{1}{12}bd^3$ will appear somewhat lengthy, and the exact value is obtained by integrating as follows.

Consider a small strip of area of breadth b and depth dy at a distance of y from the neutral axis as shown in Fig. 8.19. The moment of inertia of this strip (shown shaded) is

$$(\text{its area}) \times (y)^2 = b \times dy \times y^2 = b \times y^2 \times dy$$

The second moment of half of the rectangle is the sum of all such quantities $by^2\,dy$ between the limits of $y = 0$ and $y = \tfrac{1}{2}d$

$$I_{xx} \text{ of half rectangle} = \int_0^{d/2} by^2\,dy = \left[\frac{by^3}{3}\right]_0^{d/2} = \frac{bd^3}{24}$$

Thus,

$$I_{xx} \text{ of the complete rectangle} = \frac{bd^3}{24} \times 2 = \frac{bd^3}{12}$$

Similarly of course,

$$I_{yy} \text{ of the rectangle} = \frac{db^3}{12}$$

Figure 8.19

The values of the moments of inertia of a number of common shapes are listed below for reference (Fig. 8.20).

Rectangle, about neutral axes:

$$I_{xx} = \tfrac{1}{12}bd^3$$
$$I_{yy} = \tfrac{1}{12}db^3$$

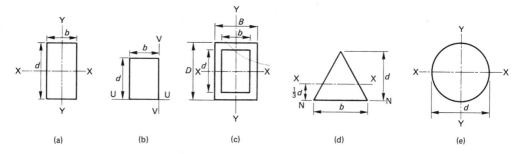

Figure 8.20

Rectangle, about one edge:

$$I_{uu} = \tfrac{1}{3}bd^3$$
$$I_{vv} = \tfrac{1}{3}db^3$$

Hollow rectangular shape:

$$I_{xx} = \tfrac{1}{12}(BD^3 - bd^3)$$
$$I_{yy} = \tfrac{1}{12}(DB^3 - db^3)$$

Triangle:

$$I_{xx} \text{ about neutral axis} = bd^3/36$$
$$I_{nn} \text{ about base} = bd^3/12$$

Circle:

$$I_{xx} = I_{yy} = \pi d^4/64$$

Figure 8.21

Figure 8.22

Principle of Parallel Axes

Fig. 8.21 shows a simple rectangular section of size $b \times d$. The I_{xx} of the rectangle is of course as previously shown $\tfrac{1}{12}bd^3$. It will be seen however that there are times when the moment of inertia of the rectangle about some other parallel axis such as Z–Z is required, and the I_{zz} will obviously be greater than the I_{xx}, since larger distances are involved in the summation AH^2.

The rule for use in these cases may be stated as follows:

To find the moment of inertia of any shape about an axis Z–Z, parallel to the neutral axis X–X and at a perpendicular distance of H away from the neutral axis, the amount AH^2 (area of shape × distance H squared) must be added to I_{xx}.

For example, in the case of the rectangle shown in Fig. 8.22,

$$I_{xx} = \tfrac{1}{12}bd^3 = \frac{150 \times 100^3}{12} = 12.5 \times 10^6 \text{ mm}^4 \text{ units}$$

The I_{zz} about the base (where $H = 50$ mm) is

$$I_{zz} = I_{xx} + AH^2 = 12.5 \times 10^6 + 15\,000 \times 50^2$$
$$= (12.5 + 37.5) \times 10^6$$
$$= 50 \times 10^6 \text{ mm}^4 \text{ units}$$

Note: This particular case could have been derived directly from the formula of $\frac{1}{3}bd^3$ given in case (b) of Fig. 8.20.

$\frac{1}{3}bd^3$ in this example is $\frac{1}{3} \times 150 \times 100^3 = 50 \times 10^6$ mm^4 units.

Example 8.4

The principle of parallel axes may be used to calculate the values of the moments of inertia of structural sections like the I-beam shown in Fig. 8.23.

The web has its centroid on the X–X axis of the beam and the I_{xx} of the web thus equals

$$\tfrac{1}{12} \times 12 \times 400^3 = 64 \times 10^6 \text{ mm}^4$$

The moment of inertia of one flange about *its own axis* (F–F) is

$$\tfrac{1}{12} \times 200 \times 24^3 = 0.23 \times 10^6 \text{ mm}^4$$

and from the principle of parallel axes, the I_{xx} of the one flange is

$$I_{xx} = 0.23 \times 10^6 + 200 \times 24 \times 212^2$$
$$= 215.96 \times 10^6 \text{ mm}^4$$

The total I_{xx} of the two flanges plus the web is

$$\text{Total } I_{xx} = (64 + 2 \times 216) \times 10^6 \text{ mm}^4$$
$$= 496 \times 10^6 \text{ mm}^4$$

An alternative method of calculating the moment of inertia about X–X is to calculate the I_{xx} of rectangle ABCD and subtract the I_{xx} of the two rectangles EFGH (Fig. 8.24). Thus

$$\text{Total } I_{xx} = \tfrac{1}{12} \times 200 \times 448^3 - \tfrac{1}{12} \times 2(94 \times 400^3)$$
$$= (1498.59 - 1002.66) \times 10^6$$
$$= 496 \times 10^6 \text{ mm}^4$$

Note: Particular care should be taken in using this method of the subtraction of moments of inertia to see that all the rectangles concerned have axis X–X as their common neutral axis.

The I_{yy} of the above beam section may most easily be calculated by adding the I_{yy} of the three rectangles of which the joist consists, as shown in Fig. 8.25, because axis Y–Y is their common neutral axis.

Figure 8.23

Figure 8.24

Figure 8.25

Figure 8.26

$$I_{yy} = 2 \times (\tfrac{1}{12} \times 24 \times 200^3) + \tfrac{1}{12} \times 400 \times 12^3$$
$$= (32.00 + 0.06) \times 10^6$$
$$= 32.06 \times 10^6 \text{ mm}^4$$

The method shown above is accurate and can be used for steel girders built up by welding together plates of rectangular cross-section. It should be noted, however, that for structural rolled steel sections the moments of inertia can be found tabulated in handbooks. Examples of such tables are given on pages 158–161 and the given moments of inertia take into account the root radius, fillets, etc.

Example 8.5

A tee section measures 140 mm × 140 mm × 20 mm as shown in Fig. 8.26. Calculate the I_{xx}.

Solution Taking moments of areas about the base to determine the position of the neutral axis X–X:

$$(2400 + 2800)\bar{x} = 2400 \times 80 + 2800 \times 10$$

$$\bar{x} = \frac{(192 + 28) \times 10^3}{5.2 \times 10^3} = 42.3 \text{ mm}$$

I_{xx} of vertical rectangle

$$\tfrac{1}{12} \times 20 \times 120^3 + 2400 \times 37.7^2$$
$$= (2.88 + 3.41) \times 10^6$$
$$= 6.29 \times 10^6 \text{ mm}^4$$

I_{xx} of horizontal rectangle

$$\tfrac{1}{12} \times 140 \times 20^3 + 2800 \times 32.2^2$$
$$= (0.09 + 2.92) \times 10^6$$
$$= 3.01 \times 10^6 \text{ mm}^4$$

Total I_{xx}

$$(6.29 + 3.01) \times 10^6 = 9.30 \times 10^6 \text{ mm}^4$$

Example 8.6

Calculate the I_{xx} and I_{yy} of the section shown in Fig. 8.27.

Solution Taking moments of area about the base to find the distance $\bar{x}$ to the neutral axis X–X:

$A_1 = 2400 \text{ mm}^2$

$A_2 = 2400 \text{ mm}^2$

$A_3 = 4800 \text{ mm}^2$

Total $A = 9600 \text{ mm}^2$

Figure 8.27

$$9600\bar{x} = 4800 \times 12 + 2400 \times 124 + 2400 \times 236$$
$$= 57\,600 + 297\,600 + 566\,400$$
$$\bar{x} = 921\,600/9600 = 96 \text{ mm}$$

The figure is divided into three rectangles as shown in Fig. 8.27 and the distances of the centroids of each from the centroid of the whole section are also given.

$$I_{xx} \text{ (top flange)} = \tfrac{1}{12} \times 100 \times 24^3 + 2400 \times 140^2$$
$$= (0.11 + 47.04) \times 10^6$$
$$= 47.15 \times 10^6 \text{ mm}^4$$

$$I_{xx} \text{ (web)} = \tfrac{1}{12} \times 12 \times 200^3 + 2400 \times 28^2$$
$$= (8.00 + 1.88) \times 10^6$$
$$= 9.88 \times 10^6 \text{ mm}^4$$

$$I_{xx} \text{ (bottom flange)} = \tfrac{1}{12} \times 200 \times 24^3 + 4800 \times 84^2$$
$$= (0.23 + 53.87) \times 10^6$$
$$= 34.10 \times 10^6 \text{ mm}^4$$

Total I_{xx} of the whole section
$$= (47.15 + 9.88 + 34.10) \times 10^6$$
$$= 91.13 \times 10^6 \text{ mm}^4$$

Total $I_{yy} = \tfrac{1}{12} \times 24 \times 100^3 + \tfrac{1}{12} \times 24 \times 200^3 + \tfrac{1}{12} \times 200 \times 12^3$
$$= (2.00 + 16.00 + 0.03) \times 10^6$$
$$= 18.03 \times 10^6 \text{ mm}^4$$

Example 8.7

Calculate the I_{xx} and I_{yy} of the channel section shown in Fig. 8.28.

Solution X–X axis:

$$I_{xx} = \tfrac{1}{12} \times 100 \times 300^3 - \tfrac{1}{12} \times 88 \times 252^3$$
$$= (225.00 - 117.36) \times 10^6$$
$$= 107.64 \times 10^6 \text{ mm}^4$$

Check by addition:

$$I_{xx} = \tfrac{1}{12} \times 12 \times 300^3 + 2 \times (\tfrac{1}{12} \times 88 \times 24^3 + 88 \times 24 \times 138^2)$$
$$= 27 \times 10^6 + 2(0.10 + 40.22) \times 10^6$$
$$= 107.64 \times 10^6 \text{ mm}^4 \quad \text{(as before)}$$

Y–Y axis:

Taking moments of areas about back of channel

Figure 8.28

$$\text{Area} \times \text{distance } \bar{x} = 300 \times 12 \times 6 + 2(24 \times 88 \times 56)$$
$$= 21\ 600 + 236\ 544$$
$$= 258\ 144 \text{ mm}^3 \text{ units}$$

$$\text{Distance } \bar{x} = \frac{258\ 144}{3600 + 2 \times 2112} = 33 \text{ mm}$$

$$I_{yy} = \tfrac{1}{12} \times 300 \times 12^3 + 3600 \times 27^2$$
$$+ 2(\tfrac{1}{12} \times 24 \times 88^3 + 2112 \times 23^2)$$
$$= [0.04 + 2.63 + 2(1.36 + 1.12)] \times 10^6$$
$$= 7.63 \times 10^6 \text{ mm}^4$$

Check by subtraction:

$$I_{yy} = (\tfrac{1}{12} \times 300 \times 100^3 + 30\ 000 \times 17^2)$$
$$- (\tfrac{1}{12} \times 252 \times 88^3 + 22\ 176 \times 23^2)$$
$$= (25.00 + 8.67 - 14.31 - 11.73) \times 10^6$$
$$= 7.63 \times 10^6 \text{ mm}^4 \quad \text{(as before)}$$

Example 8.8

Calculate the I_{xx} and I_{yy} of the compound girder shown in Fig. 8.29. The properties of the UB alone are

$$\text{Area} = 2840 \text{ mm}^2$$
$$I_{xx} = 28.67 \times 10^6 \text{ mm}^4$$
$$I_{yy} = 1.20 \times 10^6 \text{ mm}^4.$$

254 x 102 UB 22

Centroid of UB

150 mm x 18 mm plate

Figure 8.29

Solution

Note: The moment of inertia given above for the UB (without the plate) is of course about the X–X axis of the UB.

The addition of a single plate renders the compound section unsymmetrical about the X–X axis of the compound, and the AH^2 of the UB must be added to its own I_{xx}.

Taking moments about the line A–A:

$$\text{Distance } \bar{x} = \frac{2840 \times 145 + 2700 \times 9}{2840 + 2700}$$
$$= 436\ 100/5540$$
$$= 79 \text{ mm}$$

$$I_{xx} = 28.67 \times 10^6 + 2840 \times 66^2 + \tfrac{1}{12} \times 150 \times 18^3 + 2700 \times 70^2$$
$$= (28.67 + 12.37 + 0.07 + 13.23) \times 10^6$$
$$= 54.34 \times 10^6 \text{ mm}^4$$

$$I_{yy} = 1.20 \times 10^6 + \tfrac{1}{12} \times 18 \times 150^3$$
$$= (1.20 + 5.06) \times 10^6$$
$$= 6.26 \times 10^6 \text{ mm}^4$$

Figure 8.30

Example 8.9

Calculate the I_{xx} and I_{yy} of the compound girder shown in Fig. 8.30. The properties of the UB alone are:

$$\text{Area} = 2840 \text{ mm}^2$$
$$I_{xx} = 28.67 \times 10^6 \text{ mm}^4 \qquad I_{yy} = 1.20 \times 10^6 \text{ mm}^4$$

Solution

Note: The addition of a $150 \text{ mm} \times 12 \text{ mm}$ plate to each flange does not move the neutral axis of the UB from its original position, so that

$$I_{xx} = 28.67 \times 10^6 + \tfrac{1}{12} \times 150 \times 278^3 - \tfrac{1}{12} \times 150 \times 254^3$$
$$= (28.67 + 268.56 - 204.84) \times 10^6$$
$$= 92.39 \times 10^6 \text{ mm}^4$$

Check I_{xx} by addition of the I_{xx} of plates:

$$I_{xx} = 28.67 \times 10^6 + 2(\tfrac{1}{12} \times 150 \times 12^3 + 150 \times 12 \times 133^2)$$
$$= (28.67 + 2 \times 31.86) \times 10^6$$
$$= 92.39 \times 10^6 \text{ mm}^4 \quad \text{(as before)}$$
$$I_{yy} = 1.20 \times 10^6 + 2(\tfrac{1}{12} \times 12 \times 150^3)$$
$$= (1.20 + 6.75) \times 10^6$$
$$= 7.95 \times 10^6 \text{ mm}^4$$

SUMMARY

The **centre of gravity** of a body is a point in or near the body through which the resultant weight of the body acts. When dealing with areas, the terms **centroid** and **centre of area** are frequently used.

With respect to beam sections, which are plane figures without weight, the **moment of inertia** about an axis is the sum of the **second moments of area** about that axis, i.e.

$$I = \sum ay^2$$

This means that, if a section is divided into an infinite number of strips parallel to the axis in question, and the area of each strip is multiplied by the square of its distance from the axis, and then all these quantities are added together, the result is the moment of inertia of the section.

For geometrical shapes, formulae for the moments of inertia can be obtained by the aid of the calculus.

Principle of Parallel Axes: To determine the moment of inertia of any shape about an axis Z–Z parallel to another axis X–X at a perpendicular distance of H, the amount AH^2 (i.e. area of shape multiplied by the square of the distance H) must be added to the moment of inertia about X–X.

$$I_{zz} = I_{xx} + AH^2$$

Figure 8.31

EXERCISE 8

1 Fig. 8.31 shows a system of three weights connected together by rigid bars. Determine the position of the centre of gravity of the system with respect to point A, ignoring the weight of the bars.

2 Each of the shapes shown in Fig. 8.32(a), (b), (c), (d) is symmetrical about a vertical axis. Calculate the distance of the centre of gravity from the base.

3 In Fig. 8.33 (a) to (e) take point P as the intersection of the A–A and B–B axes and calculate the position of the centre of gravity with respect to these axes. ($\bar{x}$ is the distance of the c.g. from the vertical axis A–A and $\bar{y}$ is the distance of the c.g. from the horizontal axis B–B.)

Figure 8.32

Figure 8.33

All dimensions are in millimetres

4–9 Calculate the I_{xx} and the I_{yy} of the shapes shown in Figs. 8.34–8.39.

Figure 8.34

Figure 8.35

Figure 8.36

Figure 8.37

Figure 8.38

I_{xx} of UB alone = 553.53 x 10^6 mm^4
I_{yy} of UB alone = 23.92 x 10^6 mm^4

Figure 8.39

Figure 8.40

Figure 8.41

Figure 8.42

10 Two steel channels, $229\,\text{mm} \times 76\,\text{mm}$, are to be arranged as shown in Fig. 8.40, so that the I_{xx} and the I_{yy} of the compound section are equal.

The properties of one single channel are:

Area $= 3320\,\text{mm}^2$ $\qquad I_{xx} = 26.10 \times 10^6\,\text{mm}^4$

Distance $b = 20\,\text{mm}$ $\qquad I_{yy} = 15.87 \times 10^6\,\text{mm}^4$

(about axis shown dotted)

What should be the distance a?

11 The properties of a single $762 \times 267\text{UB}173$ are as follows:

Area $= 22\,050\,\text{mm}^2$

$I_{xx} = 2051.77 \times 10^6\,\text{mm}^4$

$I_{yy} = 68.48 \times 10^6\,\text{mm}^4$

Calculate the I_{xx} and I_{yy} of a compound girder which consists of two such beams at 300 mm centres and two $700\,\text{mm} \times 36\,\text{mm}$ steel plates attached to the flanges of the beams as shown in Fig. 8.41.

12–14 Determine the position of the centre of area of the shapes shown in Figs. 8.42–8.44 and calculate the values of their I_{xx}.

Figure 8.43

Figure 8.44

Figure 8.45

Figure 8.46

Figure 8.47

15 A strut is made up by inserting a 178 mm × 24 mm steel plate between two channels as shown in Fig. 8.45. The properties of a single 178 mm × 76 mm steel channel are:

$$\text{Area} = 2654 \text{ mm}^2$$
$$I_{xx} = 13.37 \times 10^6 \text{ mm}^4$$
$$I_{yy} = 1.34 \times 10^6 \text{ mm}^4$$

Calculate the values of I_{xx} and I_{yy} for the strut.

16 Calculate the I_{xx} and I_{yy} of the square section shown in Fig. 8.46.

17 A special stanchion section is built up using two UBs fixed together as shown in Fig. 8.47.

The properties of each individual UB are:

	356 × 171UB57	254 × 102UB22
Area	7220 mm²	2840 mm²
I_{xx}	160.77 × 10⁶ mm⁴	28.67 × 10⁶ mm⁴
I_{yy}	11.09 × 10⁶ mm⁴	1.20 × 10⁶ mm⁴

Calculate the dimension x to the centroid of the compound section and the values of I_{xx} and I_{yy}.

18 A timber I-beam consists of two 120 mm × 36 mm flanges and a 48 mm thick web. The overall depth of the beam is 276 mm. Determine its I_{xx} and I_{yy}.

19 A hollow circular section has an external diameter of 200 mm. The I_{xx} of the section is 52.34×10^6 mm⁴. What is the thickness of the material?

20 The overall depth of a hollow rectangular section is 300 mm and the thickness of the material is 20 mm throughout. The I_{xx} about the horizontal axis is 184.24×10^6 mm⁴. What is the overall width of the section?

21 The I_{xx} of a 352 mm deep UB is 121×10^6 mm⁴. The beam has to be increased in strength by the addition of a 220 mm wide plate to each flange so that the total I_{xx} of the compound girder is brought up to 495×10^6 mm⁴. What thickness plate will be required? Bolt or rivet holes need not be deducted.

22 The UB mentioned in Question 21 ($I_{xx} = 121 \times 10^6$ mm⁴) is to be strengthened by the addition of 30 mm thick plate to each flange so that the I_{xx} is increased to 605×10^6 mm⁴. What is the required width of the plates? Bolt or rivet holes may be ignored.

432mm x 102mm ⊏

633mm
depth of UB

23.2mm

610 x 305UB238

Figure 8.48

23 Calculate the second moment of area about the X–X (horizontal) axis of the crane gantry girder shown in Fig. 8.48.

The properties of the individual components are:

	610×305UB238	432×102⊏
Area	30 380 mm^2	8349 mm^2
I_{xx}	2075.71×10^6 mm^4	213.99×10^6 mm^4
I_{yy}	158.38×10^6 mm^4	6.29×10^6 mm^4
Web thickness	18.6 mm	12.2 mm

9 Simple Beam Design

Figure 9.1

Figure 9.2

The previous two chapters have tried to show that

1 When a beam is loaded, the beam has a bending tendency which is measured in newton millimetres (N mm), etc., and is known as the *bending moment* on the beam. There is, for every loaded beam, a certain critical point at which this bending moment has a maximum value (M_{max}).

2 The shape of the beam's cross-section has an effect upon its strength and this shape effect is measured in mm^4 units and is called the *moment of inertia* (second moment of area).

3 The material of which the beam is constructed also affects the beam's strength and this factor is measured in terms of the material's safe allowable stress (in tension or compression) and is measured in newtons per square millimetre (N/mm^2), etc.

The first of these three factors (the maximum bending moment, M_{max}) depends on the length of the beam and on the nature and disposition of the applied loads.

The second factor, i.e. the factor measuring the strength of the section by virtue of its shape, may be varied by the designer so that when the third factor, i.e. the strength of the material, is also taken into account, a beam may be designed of just sufficient strength to take the calculated bending moment.

Factor 1 may be thought of as the *punishment factor*.

A suitable combination of factors 2 and 3 may be described as the beam's *resistance factor* and, indeed, this factor is termed the *moment of resistance* M_r of the beam.

THE GENERAL THEORY OF BENDING

It has been seen that, in general, most beams tend to bend in the form shown in Fig. 9.1, and if X–X is the neutral axis of such a beam, the fibres above X–X are stressed in compression and those below X–X in tension.

Furthermore, fibres far away from the neutral axis are stressed more heavily than those near to the neutral axis. Fibres lying on the neutral axis are neither in tension or compression, and are, in fact, quite unstressed.

The distribution of stress across a plane section such as A–A in Fig. 9.1 may thus be seen to vary from zero at the neutral axis to maximum tension and maximum compression at the extreme fibres, as shown in the stress distribution diagram (Fig. 9.2).

The stress varies directly as the distance from the neutral axis, i.e. the stress at C is twice that at B.

You should appreciate now a statement made in earlier chapters—that material far away from the neutral axis is more useful than material near to the neutral axis. You should also appreciate why the Universal Beam (UB) is such a popular beam section; the mass of its material (i.e. the two flanges) is positioned as far as possible from the centroid, that is, from the neutral axis (NA).

A simple analogy will help to show how the compression and tension generated within the beam combine to resist the external bending moment.

Consider, for example, the cantilever shown in Fig. 9.3, and let a load of W kN be suspended from the free end B, the self-weight of the beam being neglected. If the beam were to be sawn through at section C–C then the portion to the right of the cut would fall, but it could be prevented from falling by an arrangement as shown in Fig. 9.4.

A load of W kN, represented by the weight W, is attached to a wire which, in turn, passes over a pulley P and is connected to the small portion of cantilever. (The weight of the cantilever itself is ignored for the purpose of this discussion.)

However, although this arrangement would prevent the small portion from falling, the portion would still not be in equilibrium, for it would now twist or rotate as shown by the broken lines. This rotation could also be prevented by a small length of chain D–D and a steel block at S–S, as shown in Fig. 9.5.

This is because the chain exerts a pull (tension) on the portion of cantilever to the right, and the steel block exerts a push (compression) on the portion.

These forces T and C are equal to each other, and they form an *internal couple* whose moment is equal to and resists the bending moment Wx. This couple provides the moment of resistance of the section which equals $T \times$ lever arm or $C \times$ lever arm. In a real beam these forces T and C are, of course, provided by the actual tension and compression in the many fibres of the cross-section.

Figure 9.3

Figure 9.4

Figure 9.5

Similarly, in a real beam, the force W which prevents the portion from falling is provided by the shear stress generated within the beam itself.

A cantilever has been chosen to illustrate the nature of these internal forces and moments, but the principle remains the same in the case of the simply supported beam, except that the tension will now occur in the layers below the NA and compression in those above the NA.

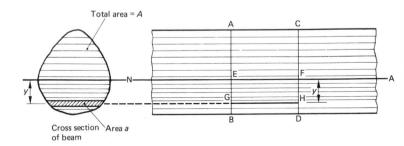

Figure 9.6

Fig. 9.6 shows a short portion of beam (of any shape) before loading, with two plane sections A–B and C–D taken very close together. The layer E–F lies on the NA and layer G–H is one at a distance of y below the NA and having a small cross-sectional area of a.

Fig. 9.7 shows the same length of beam after bending. The sections A–B and C–D remain straight but are rotated to the positions shown, A_1–B_1 and C_1–D_1. The small portion has bent through an angle of ϕ radians, at a radius of curvature (measured to the NA) of R.

The fibre AC has decreased in length to A_1–C_1, whilst the fibre BD has increased in length to B_1–D_1. G–H has also increased to G_1–H_1 but the increase is less than in the case of the fibre B–D. The fibre E–F, being on the NA, neither increases nor decreases and E–F = E_1–F_1.

This straining (alteration in length) of the various fibres in the depth of the section is shown in Fig. 9.8(*b*), and as stress and strain are proportional to each other (stress/strain = E, see Chapter 6) a similar diagram, 9.8(*c*), shows the distribution of stress over the depth of the section.

Referring again to Fig. 9.7, the fibre G–H increases in length to G_1–H_1. Therefore

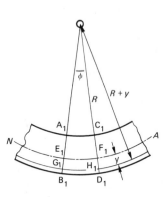

Figure 9.7

$$\text{Strain in fibre G–H} = \frac{\text{increase in length}}{\text{original length}}$$

$$= \frac{(G_1\text{–}H_1) - (G\text{–}H)}{G\text{–}H}$$

Figure 9.8

But length of arc = radius × angle in radians, therefore

$$G_1\text{-}H_1 = (R + y)\phi$$

Now E_1-F_1 remains the same length and thus E_1-$F_1 = $ G-H. Also E_1-$F_1 = R\phi$.

Thus

$$\frac{(G_1\text{-}H_1) - (G\text{-}H)}{G\text{-}H} = \text{strain in fibres G-H} = \frac{(R + y)\phi - R\phi}{R\phi}$$

$$= \frac{R\phi + y\phi - R\phi}{R\phi} = \frac{y}{R} \qquad (1)$$

and, since stress/strain $= E$, then stress $=$ strain $\times E$, therefore,

$$\text{Stress in fibres G-H} = f = \frac{y}{R} \times E = \frac{Ey}{R} \qquad (2)$$

i.e.

$$\frac{f}{y} = \frac{E}{R} \qquad (3)$$

The total load over the area $a = $ stress on the area × the area

$$= \frac{Ey}{R} \times a = \frac{Eay}{R}$$

The first moment of the load about the NA is

$$\text{Load} \times y = \frac{Eay^2}{R}$$

The summation of all such moments over the total area A (i.e. the sum of the moments of all such areas as a), including areas above the neutral axis as well as below, is

$$\sum \frac{Eay^2}{R}$$

and this is the total moment of resistance of the section.

But E and R are constants (i.e. only a and y^2 vary) and, since in design the moment of resistance is made equal to the bending moment (usually the maximum bending moment, M_{max}),

$$M_{max} = M_r = \frac{E}{R} \sum ay^2$$

i.e.

Safe bending moment $= \dfrac{E}{R} \times$ the sum of all the ay^2

But as was already stated in Chapter 8, the sum of all the ay^2, i.e. $\sum ay^2$, is the moment of inertia or second moment of area of the shape, I. So it may now be written

$$M_r = \frac{EI}{R} \tag{4}$$

Referring again to equation (3), $f/y = E/R$ and substituting f/y for E/R in equation (4),

$$M_r = \frac{fI}{y} = f\left(\frac{I}{y}\right) \tag{5}$$

Finally, the main formulae arising from this theory may be stated as follows:

$$\frac{M_r}{I} = \frac{f}{y} = \frac{E}{R} \tag{6}$$

and the most important portion consists of the first two terms

$$\frac{M_r}{I} = \frac{f}{y} \quad \text{or} \quad M_r = f\left(\frac{I}{y}\right) \quad \text{i.e. equation (5)}$$

In the above formula f is the permissible bending stress for the material of the beam. For the most commonly used steel (Grade 43), f is taken as 165 N/mm^2, whereas the value of f for timber ranges between 3 N/mm^2 and 12 N/mm^2 depending on the species.

These values, however, apply only in those cases where the compression flange of the beam is adequately restrained laterally to prevent sideways buckling. For beams which are not restrained laterally, the above values have to be appropriately reduced. This is outside the scope of this textbook and it can be assumed that all the beams dealt with in this chapter are adequately supported laterally.

SECTION MODULUS: 1) SYMMETRICAL SECTIONS

In equation (5) the factor I/y is called the **section modulus** (or modulus of section) and is usually denoted by the letter Z,

$$\frac{I}{y} = Z$$

Where a beam is symmetrical about its X–X axis, as in the case of those shown in Fig. 9.9, the X–X axis is also the neutral axis (NA) of the beam section, and so the distance y to the top flange is the same as that to the bottom flange, hence

$$Z = \frac{I_{xx}}{y} = \frac{I_{xx}}{\frac{1}{2}d}$$

Now, for a rectangular section, $I_{xx} = \frac{1}{12}bd^3$, and so

$$Z = \frac{I}{y} = \frac{bd^3}{12} \div \frac{d}{2} = \frac{bd^2}{6}$$

In the case of rolled steel sections the values of Z (i.e. I/y) are normally taken directly from tables similar to those on pages 159 and 161 (Elastic Modulus column) but the following examples include some in which the values are calculated.

Figure 9.9

Example 9.1

A timber beam of rectangular cross-section is 150 mm wide and 300 mm deep. The maximum allowable bending stress in tension and compression must not exceed 6 N/mm². What maximum bending moment in N mm can the beam safely carry?

Solution

Section modulus $Z = \frac{1}{6}bd^2 = \frac{1}{6} \times 150 \times 300^2$
$$= 2.25 \times 10^6 \text{ mm}^3 \text{ units}$$

Safe allowable bending moment $M_{max} = fZ = 6 \times 2.25 \times 10^6$
$$= 13.5 \times 10^6 \text{ N mm}$$

Example 9.2

A timber beam has to support loading which will cause a maximum bending moment of 25×10^6 N mm. The safe bending stress must not exceed 7 N/mm². What section modulus will be required in choosing a suitable size section?

Solution

$M_r = fZ$ and, therefore, the required $Z = M_r/f$, i.e.

$Z = 25 \times 10^6/7 = 3.57 \times 10^6$ mm³

Example 9.3

A timber beam is required to span 4 m carrying a total uniform load (inclusive of the beam's self-weight) of 40 kN. The safe allowable bending stress is 8 N/mm². Choose a suitable depth for the beam if the width is to be 120 mm.

Solution

$$M_{max} = \tfrac{1}{8}Wl = \tfrac{1}{8} \times 40 \times 4 \times 10^6 = 20 \times 10^6 \text{ N mm}$$

but $M_{max} = M_r = f \times Z$, and it follows that

$$\text{required } Z = \frac{M_{max}}{f} = \frac{20 \times 10^6}{8} = 2.5 \times 10^6 \text{ mm}^3.$$

The Z of the section is $\tfrac{1}{6}bd^2$, where b is given as 120 mm, therefore

$$\tfrac{1}{6} \times 120 \times d^2 = 2.5 \times 10^6 \text{ mm}^3$$

Thus $d^2 = 2.5 \times 10^6 \times 6/120$

$d = \sqrt{125\,000} = 353$ mm, say 360 mm

Example 9.4

The section of floor shown in Fig. 9.10 is to be carried by 125 mm × 75 mm timber joists spanning the 3 m length. The bending stress must not exceed 4.6 N/mm² and the total inclusive load per m² of floor is estimated to be 2.0 kN. At what cross-centres x in mm must the timber beams be fixed?

Solution Total load carried by *one* timber joist is

$3 \times x \times 2.0 = 6x$ kN

Z of joist $= \tfrac{1}{6}bd^2 = \tfrac{1}{6} \times 75 \times 125^2 = 195 \times 10^3$ mm³

Moment of resistance of one joist = bending moment on one joist

$$M_r = fZ = 4.6 \times 195 \times 10^3 = 0.9 \times 10^6 \text{ N mm}$$

Thus

$$M_{max} = \tfrac{1}{8}Wl = \tfrac{1}{8} \times 6x \times 3 \times 10^6 = 0.9 \times 10^6 \text{ N mm}$$

and

$$x = \frac{0.9 \times 10^6 \times 8}{6 \times 3 \times 10^6} = 0.4 \text{ m}$$

Figure 9.10

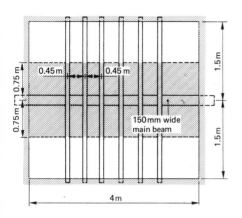

Figure 9.11

Example 9.5

If the floor mentioned in Example 9.4 has its 125 mm × 75 mm timber joists at 450 mm centres and the span of the timber joists is halved by the introduction of a main timber beam 150 mm wide, as shown in Fig. 9.11, (*a*) what load in kN/m² will the floor now safely carry? (*b*) what will be the required depth of the 150 mm wide main timber beam if $f = 5.25\ \text{N/mm}^2$?

Solution Let w be safe load in kN/m². Then load carried by one secondary timber joist is

$$1.5 \times 0.45 \times w = 0.675w\ \text{kN}$$

M_r of the joists is unchanged at $0.9 \times 10^6\ \text{N mm}$, therefore

$$M_{max} = \tfrac{1}{8} \times 0.675w \times 1.5 \times 10^6 = 0.9 \times 10^6\ \text{N mm}$$

and

$$w = \frac{0.9 \times 10^6 \times 8}{0.675 \times 1.5 \times 10^6} = 7\ \text{kN/m}^2$$

Total load on the main beam

Each secondary timber joist transfers half of its load to the main beam in the form of an end reaction, and the other half to the wall.

Thus, the total area of floor load carried by the main beam is the area shown shaded in Fig. 9.11.

The total load carried by the main beam is thus $4.0 \times 1.5 \times 7 = 42\ \text{kN}$ and, although this in fact consists of a large number of small point loads from the secondary beams, it is normal practice to treat this in design as a uniformly distributed load. Therefore

$$M_{max}\ \text{on main beam} = \tfrac{1}{8} \times 42 \times 4 \times 10^6 = 21 \times 10^6\ \text{N mm}$$

and

$$Z\ \text{required} = \frac{M_r}{f} = \frac{21 \times 10^6}{5.25} = 4 \times 10^6\ \text{mm}^3$$

but for a rectangular section, $Z = \tfrac{1}{6}bd^2$, so

$$d = \sqrt{\frac{4 \times 10^6 \times 6}{150}} = 400\ \text{mm}$$

Example 9.6

A welded steel girder is made up of plates as shown in Fig. 9.12.

What safe, uniformly distributed load can this girder carry

Figure 9.12

Brickwork at 20 kN/m³
= 0.22 x 2.6 x 5.5 x 20 = 63 kN

5.5 m

Beam, say, 2 kN self-weight

Figure 9.13

on a simply supported span of 4.0 m if the permissible bending stress is limited to 165 N/mm²?

Solution

$$I_{xx} \text{ of section} = \tfrac{1}{12} \times 120 \times 300^3 - \tfrac{1}{12} \times 108 \times 250^3$$
$$= 270 \times 10^6 - 140.625 \times 10^6$$
$$= 129 \times 10^6 \text{ mm}^4$$

$$Z_{xx} \text{ of section} = I/y = 129 \times 10^6/150 = 0.86 \times 10^6 \text{ mm}^3$$

Hence

$$M_r = 165 \times 0.86 \times 10^6 = 142 \times 10^6 \text{ N mm}$$

Let W be the total safe, uniformly distributed load the girder can carry (in kN), then

$$\tfrac{1}{8}W \times 4 \times 10^6 = 142 \times 10^6 \text{ N mm}$$
$$W = 142 \times 8/4 = 284 \text{ kN}$$

Example 9.7

A steel beam is required to span 5.5 m between centres of simple supports carrying a 220 mm thick brick wall as detailed in Fig. 9.13.

Choose from the tables of properties (pages 158–161) a suitable beam section given that the permissible stress in bending is 165 mm².

Solution

Total uniformly distributed load $W = 63 + 2 = 65$ kN

Therefore,

$$M_{max} = \tfrac{1}{8} \times 65 \times 5.5 \times 10^6 = 44.69 \times 10^6 \text{ N mm}$$

and

$$\text{required } Z = 44.69 \times 10^6/165 = 270\,800 \text{ mm}^3$$
$$\text{or } 270.8 \text{ cm}^3$$
$$\text{(as shown in tables)}$$

A suitable section would be the 305×102UB25 (i.e. 305 mm × 102 mm @ 25 kg/m Universal Beam) with an actual value of Z at 287.9 cm³.

The 203×133UB30 with a Z value of 279.3 cm³ is nearer to the required 270.5 cm³, but the former section was chosen because of its smaller weight.

Example 9.8

Fig. 9.14 shows a portion of floor of a steel-framed building. The floor slab, which weighs 3 kN/m², spans in the direction of the arrows, carrying a super (live) load of 5 kN/m² and transferring it to the secondary beams marked A. These in turn pass the loads to the main beams marked B and to the columns.

Choose suitable sizes for beams A and B, using a permissible bending stress of 165 N/mm².

Solution Secondary beam, A:

Total load $W = 3.8 \times 7.5 \times (3 + 5) = 228$ kN

$M_{max} = \frac{1}{8} \times 228 \times 7.5 \times 10^6 = 213.75 \times 10^6$ N mm

and

$$\text{required } Z = \frac{213.75 \times 10^6}{165} = 1.295 \times 10^6 \text{ mm}^3$$

$$\text{or } 1295 \text{ cm}^3$$

Use 457×191UB67 with a Z value of 1296 cm³.

Figure 9.14

Main beam, B:
Reaction from beams A at mid-span = 228 kN (point load).
Assumed weight of beam (udl), say, 3 kN/m

$$M_{max} = (\tfrac{1}{4} \times 228 + \tfrac{1}{8} \times 3 \times 7.6) \times 7.6 \times 10^6$$
$$= (57 + 2.85) \times 7.6 \times 10^6$$
$$= 454.86 \times 10^6 \text{ N mm}$$

and

$$\text{required } Z = \frac{454.86 \times 10^6}{165} = 2.757 \times 10^6 \text{ mm}^3$$

$$\text{or } 2757 \text{ cm}^3$$

Use 610×229UB113 with a Z value of 2879 cm³.

Example 9.9

The timber box beam shown in Fig. 9.15 spans 6 m and carries a uniformly distributed load of 2 kN/m and a centrally placed point load of 3 kN. Determine the maximum stresses due to bending.

Solution Because of the symmetry of the loading, the maximum bending moment due to the combined action of the two loads will occur at mid-span.

$$M_{max} = (\tfrac{1}{8} \times 2 \times 6 + \tfrac{1}{4} \times 3) \times 6 \times 10^6 = 13.5 \times 10^6 \text{ N mm}$$

$$I_{xx} = \tfrac{1}{12} \times 120 \times 450^3 - \tfrac{1}{12} \times 96 \times 350^3$$

$$= (911.25 - 343.00) \times 10^6 = 568.25 \times 10^6 \text{ mm}^4$$

Figure 9.15

$$Z_{xx} = \frac{568.25 \times 10^6}{225} = 2.525 \times 10^6 \text{ mm}^3$$

Now, from $M = f \times Z$, $f = M/Z$, therefore

$$f = \frac{13.5 \times 10^6}{2.525 \times 10^6} = 5.35 \text{ N/mm}^2$$

SECTION MODULUS: 2) NONSYMMETRICAL SECTIONS

Figure 9.16

Fig. 9.16 shows two examples of beam sections which are not symmetrical about the X–X axis. In such cases the distance y between the neutral axis and the top and bottom fibres will have two values, y_1 and y_2, and, consequently, there will also be two values of Z_{xx}—one obtained by dividing I_{xx} by the distance y_1, and the other obtained by dividing I_{xx} by the value of y_2.

If in these unsymmetrical shapes the permissible stress is the same for both top and bottom flanges, as it is, for example, in the case of steel, then

Safe bending moment $= f \times$ least Z

If, as in the case of the now obsolete cast iron joist shown in Fig. 9.16, the permissible stresses in tension and compression are not the same, then the safe bending moment is the lesser of the following:

$$M_{re} = \frac{I}{y_1} \times \text{compression stress}$$

$$M_{rt} = \frac{I}{y_2} \times \text{tension stress}$$

Example 9.10

A 254×102UB22 has one 200 mm × 12 mm steel plate welded to the top flange only as shown in Fig. 9.17. The area of the Universal Beam alone is 2840 mm² and $I_{xx} = 28.67 \times 10^6$ mm⁴. What safe uniform load can such a beam carry on a 5.0 m simply supported span if $f = 165$ N/mm² in tension or compression?

Solution Taking moments about line A–A to find the neutral axis which passes through the centroid of the section:

$$(2840 + 2400)x = 2840 \times 127 + 2400 \times 260 = 984\,680$$

$$x = 984\,680/5240 = 188 \text{ mm}$$

Figure 9.17

Applying the Principle of Parallel Axes (Chapter 8) to determine the second moment of area of the section about NA:

$$I_{NA} = 28.67 \times 10^6 + 2840 \times 61^2 + \tfrac{1}{12} \times 200 \times 12^3 + 2400 \times 72^2$$
$$= (28.67 + 10.57 + 0.03 + 12.44) \times 10^6$$
$$= 51.71 \times 10^6 \text{ mm}^4$$
$$Z_{top} = 51.71 \times 10^6 / 78 = 663\,000 \text{ mm}^3$$
$$Z_{bot} = 51.71 \times 10^6 / 188 = 275\,000 \text{ mm}^3$$

Therefore

Critical $M_r = 165 \times 275\,000 = 45.38 \times 10^6$ N mm

For a udl, $M_{max} = \tfrac{1}{8} Wl$

$$W = 45.38 \times 8/5 = 72.6 \text{ kN}$$

This load will cause the bottom fibres of the beam to be stressed in tension at 165 N/mm², while the maximum compressive stress in the top fibres will be much less, namely

$$165 \times \frac{275\,000}{663\,000} \quad \text{or, more simply,}$$

$$165 \times \frac{78}{188} = 68.5 \text{ N/mm}^2$$

Example 9.11

Fig. 9.18 shows an old type cast iron joist with a tension flange of 9600 mm², a compression flange of 2880 mm² and a web of 7200 mm².

The safe stress in compression is 5 N/mm², and in tension 2.5 N/mm². *a*) What is the safe bending moment for the section? *b*) What safe uniform load will the beam carry on a 4.8 m span?

Solution Taking moments about the lower edge A–A to find the neutral axis,

$$2\,880 \times 360 = 1\,036\,800$$
$$7\,200 \times 198 = 1\,425\,600$$
$$9\,600 \times \ 24 = \ \ 230\,400$$
$$19\,680 \qquad = 2\,692\,800$$

Therefore $x = 2\,692\,800 / 19\,680 = 137$ mm

Total I_{NA} is

$$\tfrac{1}{12} \times 120 \times 24^3 + 2880 \times 223^2 = 143.36 \times 10^6$$
$$\tfrac{1}{12} \times 24 \times 300^3 + 7200 \times 61^2 \ = \ 80.79 \times 10^6$$
$$\tfrac{1}{12} \times 200 \times 48^3 + 960 \times 113^2 \ = \underline{124.42 \times 10^6}$$
$$348.57 \times 10^6 \text{ mm}^4$$

Dimensions are in mm

Figure 9.18

Safe bending moment in tension = tension $f \times$ tension Z

$$= 2.5 \times \frac{348.57 \times 10^6}{137}$$

$$= 6.36 \times 10^6 \text{ N mm}$$

Safe bending moment in compression

$$= \text{compression } f \times \text{compression in } Z$$

$$= 5.0 \times \frac{348.57 \times 10^6}{235}$$

$$= 7.42 \times 10^6 \text{ N mm}$$

a) Thus tension is the critical stress, and

Safe bending moment $= 6.36 \times 10^6 \text{ N mm}$

b) $\frac{1}{8}Wl = 6.36 \times 10^6 \text{ N mm}$

$W = 6.36 \times 8/4.8 = 10.6 \text{ kN}$

Example 9.12

A timber joist has a 48 mm thick flange of the same timber fixed to it as shown below in Fig. 9.19, so that the resulting section may be considered to act as a tee beam.

Calculate the safe distributed load on a span of 3.6 m with an extreme fibre stress of 7 N/mm².

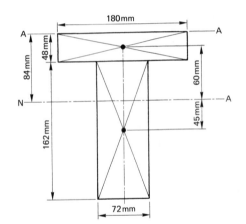

Figure 9.19

Solution Taking moments about line A–A to find the NA:

$$8640 \times 24 \qquad\qquad = \quad 207\,360$$
$$\underline{11\,664 \times (48 + \tfrac{1}{2}162)} = \underline{1\,504\,656}$$
$$20\,304 \qquad\qquad\qquad 1\,712\,016$$

Therefore $\quad x = 1\,712\,016/20\,304 = 84 \text{ mm}$

Total I_{NA} is

$$\tfrac{1}{12} \times 180 \times 48^3 + 8640 \times 60^2 + \tfrac{1}{12} \times 72 \times 162^3 + 11\,664 \times 45^2$$
$$= (1.66 + 31.10 + 25.51 + 23.63) \times 10^6$$
$$= 81.9 \times 10^6 \text{ mm}^4$$

Least $Z_{xx} = \dfrac{I_{xx}}{y_{tension}} = \dfrac{81.9 \times 10^6}{210 - 84} = 650\,000 \text{ mm}^3$

Therefore

$$M_r = fZ = 7 \times 650\,000 = 4.55 \times 10^6 \text{ N mm}$$

If W = safe distributed load in kN,

$$\tfrac{1}{8}Wl = 4.55 \times 10^6 \text{ N mm}$$

$$W = \frac{4.55 \times 10^6 \times 8}{3600} = 10\,000 \text{ N} = 10 \text{ kN}$$

Example 9.13

A timber cantilever beam projects 2 m and carries a 6 kN point load at the free end. The beam is 150 mm wide throughout, but varies in depth from 150 mm to 250 mm, as shown in Fig. 9.20. Calculate the stress in the extreme fibres *a*) at the support, *b*) at a point 1 m from the support. Ignore the weight of the beam.

Figure 9.20

Solution

a) To find stress at the support:

$$M_{max} = 6 \times 2 \times 10^6 = 12 \times 10^6 \text{ N mm}$$

$$Z \text{ at point A} = \tfrac{1}{6}bd^2 = \tfrac{1}{6} \times 150 \times 250^2 = 1.56 \times 10^6 \text{ mm}^3$$

$$\text{Stress } f \text{ at A} = \frac{M}{Z} = \frac{12 \times 10^6}{1.56 \times 10^6} = 7.68 \text{ N/mm}^2$$

b) To find stress at point B

$$M_B = 6 \times 1 \times 10^6 = 6 \times 10^6 \text{ N mm}$$

$$Z \text{ at point B} = \tfrac{1}{6}bd^2 = \tfrac{1}{6} \times 150 \times 200^2 = 1.0 \times 10^6 \text{ mm}^3$$

$$\text{Stress } f \text{ at B} = \frac{M}{Z} = \frac{6 \times 10^6}{1 \times 10^6} = 6.0 \text{ N/mm}^2$$

Example 9.14

A hollow steel pipe of 150 mm external and 100 mm internal diameter, as shown in Fig. 9.21 is to span between two buildings. What is the greatest permissible span in metres if the stresses in tension and compression must not exceed 150 N/mm²?

Note: The unit weight of steel is 78 kN/m³

Solution

$$\text{Area} = \tfrac{1}{4}\pi(150^2 - 100^2) = 0.7854 \times (22\,500 - 10\,000)$$
$$= 9.8 \times 10^3 \text{ mm}^2$$

Figure 9.21

If l = length of span in millimetres, then

Total volume of pipe = $l \times 9800 \text{ mm}^3$

Total weight of pipe = $\dfrac{l \times 9800 \times 78\,000}{1\,000\,000\,000} = 0.77 \times l \text{ N}.$

But I_{xx} of pipe = $\frac{1}{64}\pi(150^4 - 100^4) = \frac{1}{64} \times \pi \times 406.25 \times 10^6 = 19.94 \times 10^6 \text{ mm}^4$, therefore

Z_{xx} of pipe = $I/y = 19.94 \times 10^6/75 = 0.27 \times 10^6 \text{ mm}^3$

Permissible $M_{max} = M_r = \frac{1}{8}Wl = fZ = 150 \times 0.27 \times 10^6$
$$= 40.5 \times 10^6 \text{ N mm}$$

therefore

$\frac{1}{8} \times 0.77l \times l = 40.5 \times 10^6 \text{ N mm}$

$l = \sqrt{\dfrac{40.5 \times 10^6 \times 8}{0.77 \times 10^6}} = 20.5 \text{ m}$

SUMMARY

Design of Timber and Steel Beams

$M_r = f \times I/y = fZ$
  = moment of resistance
  = maximum bending moment, M_{max}
    in N mm, kN m etc.
f = maximum permissible bending stress in N/mm²
I = moment of inertia of the section about the axis of bending (mm⁴)
y = the distance from the neutral axis to the extreme beam fibres.
Z = section modulus (mm³) = I/y.

For rectangular cross-section beams, $Z = \frac{1}{6}bd^2$.
For circular cross-section beams, $Z = \pi D^3/32$.
For rolled steel sections, values of Z can be found from Tables 9.2A and 9.2B (pp. 158–161) under Elastic Modulus.
For sections built up with rectangular plates, values of Z can be calculated from first principles, having first obtained the moments of inertia (see Chapter 8).

Figure 9.22

Figure 9.23

Figure 9.24

Figure 9.25

EXERCISE 9

1 A 250 mm × 75 mm timber beam with its longer edge vertical spans 2 m between simple supports. What safe uniformly distributed load W can the beam carry if the permissible bending stress is 8 N/mm²?

2 A timber joist 75 mm wide has to carry a uniform load of 10 kN on a span of 4 m. The bending stress is to be 6 N/mm². What depth should the joist be?

3 A steel girder, as shown in Fig. 9.22 spans 3 m carrying a uniform load over the entire span. The bending stress is to be 150 N/mm². What may be the value of the uniform load?

4 The properties of a 356 × 171 UB45 are as follows:

$$I_{xx} = 120.91 \times 10^6 \text{ mm}^4 \qquad I_{yy} = 8.12 \times 10^6 \text{ mm}^4$$

What safe uniform load will the beam carry on a span of 4 m, if the stress is to be 125 N/mm²?

5 The Universal Beam used in Question 4, now spans 4.8 m carrying a uniform load of 170 kN. What is the maximum bending stress in N/mm²?

6 Two 254 mm × 89 mm steel channels, arranged back to back with 20 mm space between them, act as a beam on a span of 4.8 m. Each channel has a section modulus (Z) about axis x–x of 350 000 mm³. Calculate the maximum point load that the beam can carry at mid-span if the safe stress is 165 N/mm² and the beam's self weight is ignored.

7 A timber tee-beam is formed by rigidly fixing together two joists, as shown in Fig. 9.23. The resulting section is used (with the 216 mm rectangle placed vertically) as a beam spanning 3.6 m between simple supports. Calculate the safe uniformly distributed load if the permissible fibre stress is 5.6 N/mm².

8 A 152 mm × 76 mm @ 19 kg/m steel tee section, as shown in Fig. 9.24, may be stressed to not more than 155 N/mm². What safe inclusive uniform load can the section carry as a beam spanning 2.0 m between simple supports?

9 Timber floor joists 200 mm × 75 mm at 300 mm centres span 3 m between centres of simple supports (Fig. 9.25). What safe inclusive load, in kN/m², may the floor carry if the timber stress is not to exceed 6.0 N/mm²?

10 A small floor 4.8 m × 4.2 m is to be supported by one main beam and 150 mm × 48 mm timbers spanning between the wall and the beam, as shown in Fig. 9.26. *a*) Calculate the safe inclusive floor load if the stress in the timber has a maximum value of 7.0 N/mm². *b*) Choose a suitable section modulus for the main steel beam if the stress is not to exceed 165 N/mm². Ignore the weights of the joists and the beam.

2.4m   2.4m

150mm x 48mm joists at 360mm c/c

150mm x 48mm joists at 360mm c/c

4.2 m

Figure 9.26

100mm

24mm

24mm

360mm

N ———————————— A

x

36mm

200mm

Figure 9.27

11 *a*) Calculate the dimension *x* to the centroid of the section shown in Fig. 9.27; *b*) Determine I_{xx} and the two values of Z_{xx} for the section; *c*) What safe inclusive uniformly distributed load can a beam of this section carry on a span of 3.6 m if the tension stress must not exceed 20 N/mm² and the compression stress 100 N/mm²?

12 How deep would a 150 mm wide timber beam need to be to carry the same load as the beam investigated in Question 11 if the maximum flexural stress equals 8 N/mm²?

13 Choose suitable section moduli and select appropriate UBs for the conditions shown in Fig. 9.28 (*a*) to (*f*). $f = 165$ N/mm².

14 A UB 630 mm deep has a section modulus (Z_{xx}) of 4×10^6 mm³. The beam spans 10 m carrying an inclusive uniform load of 100 kN and a central point load of 50 kN; *a*) calculate the maximum stress due to bending; *b*) what is the intensity of flexural stress at a point 150 mm above the neutral axis and 3.0 m from l.h. reaction?

15 A steel tank 1.8 m × 1.5 m × 1.2 m weighs 15 kN empty and is supported, as shown in Fig. 9.29, by two steel beams weighing 10 kN each. Choose a suitable section for the steel beams if the tank may be filled with water weighing 10 kN/m³. The permissible bending stress on the section is 165 N/mm².

16 Choose suitable timber joists and calculate sizes for UBs A and B for the floor shown in Fig. 9.30 if the inclusive floor loading is 8.4 kN/m².

Permissible timber stress = 7.2 N/mm²

Permissible steel stress = 165 N/mm²

17 The symmetrically loaded beam, shown in Fig. 9.31, carries three loads, and the internal span *l* has to be such that the negative bending moment at each support equals the positive bending moment at C. What is the span *l*? If each load *W* is 100 kN, choose a suitable UB ($f = 165$ N/mm²).

Figure 9.28

40 kN

70 kN

1.4m

4.2m

(a)

60 kN   30 kN
2m

4m   2m

(b)

80 kN
30 kN/m   30 kN/m

2m 3m   3m 2m
6m

(c)

30 kN/m   40 kN
3m   2m   1m

60 kN/m

6m

(d)

30 kN/m   30 kN/m
90 kN/m

3m   2m   1m
6m

(e)

40 kN
30 kN/m   30 kN/m

1.2m   1.0m   1.2m   0.6m 1.0m
5m

(f)

Figure 9.29

Figure 9.30

Figure 9.31

18 A 610×305UB149, 610 mm deep ($I_{xx} = 1.25 \times 10^9$ mm^4) is simply supported on a span of l m. What will be the maximum permissible span if the stress in the beam under its own weight reaches 22 N/mm^2?

19 The plan of a floor of a steel-framed building is as shown in Fig. 9.32. Reinforced concrete slabs spanning as indicated ($\leftrightarrow$) are supported by steel beams AB. Each beam AB carries a stanchion at C, and the point load from each stanchion is 90 kN. The total inclusive loading will be 9 kN/m^2, which may be assumed to include the weight of the beams. Calculate and select a suitable beam AB for the floor, using a safe bending stress of 160 N/mm^2.

20 A steel beam carries loads, as shown in Fig. 9.33. Calculate the position and amount of the maximum bending moment and draw the Shear Force and Bending Moment Diagrams. Choose a suitable steel section for the beam, using a safe bending stress of 150 N/mm^2.

21 Two 254 mm $\times$ 89 mm steel channels, arranged back to back with a 12 mm space between them, act as a simply supported beam on a span of 4.25 m. Each has a Z_{xx} of 0.35×10^6 mm^3 and a web thickness of 9.1 mm. The safe stress is 155 N/mm^2. *a*) Ignoring the self-weight of the steel channels, calculate the maximum safe central point load which can be suspended from the channels. *b*) If this load is suspended by a short steel flat member 12 mm thick, fitted and riveted between the channels, calculate the necessary width for the flat member assuming two rows of rivets at 18 mm diameter. The safe stress in tension = 150 N/mm^2.

22 A rectangular timber beam 300 mm deep and 250 mm wide, freely supported on a span of 6 m carries a uniform load of 3 kN and a triangular load of 12 kN as shown in Fig. 9.34.

What is the greatest central point load that can be added to this beam if the maximum bending stress is 8 N/mm^2?

Figure 9.33

Figure 9.34

Figure 9.32

Table 9.1 Universal beams: safe loads for Grade 43 steel (from BCSA Structural Steelwork Handbook)

Serial size	Mass per metre	Safe distributed loads in kilonewtons for spans in metres and deflection coefficients													Critical span Lc
		4.00	5.00	6.00	7.00	8.00	9.00	10.00	11.00	12.00	13.00	14.00	15.00	16.00	
mm	kg	28.00	17.92	12.44	9.143	7.000	5.531	4.480	3.702	3.111	2.651	2.286	1.991	1.750	m
914×419	388		†3958	3429	2939	2572	2286	2057	1870	1714	1583	1470	1372	1286	8.44
	343		†3536	3012	2582	2259	2008	1807	1643	1506	1390	1291	1205	1130	8.08
914×305	289	3588	2871	2392	2051	1794	1595	1435	1305	1196	1104	1025	957	897	5.60
	253	3132	2505	2088	1789	1566	1392	1253	1139	1044	964	895	835	783	5.39
	224	2720	2176	1813	1554	1360	1209	1088	989	907	837	777	725	680	5.14
	201	2373	1899	1582	1356	1187	1055	949	863	791	730	678	633	593	4.86
838×292	226	2630	2104	1754	1503	1315	1169	1052	957	877	809	752	701	658	5.29
	194	2189	1751	1459	1251	1094	973	876	796	730	674	625	584	547	4.94
	176	1940	1552	1293	1109	970	862	776	705	647	597	554	517	485	4.72
762×267	197	2054	1643	1369	1173	1027	913	821	747	685	632	587	548	513	4.86
	173	1773	1419	1182	1013	887	788	709	645	591	546	507	473	443	4.61
	147	1475	1180	984	843	738	656	590	537	492	454	422	393	369	4.33
686×254	170	1618	1294	1078	924	809	719	647	588	539	498	462	431	404	4.73
	152	1440	1152	960	823	720	640	576	524	480	443	411	384	360	4.57
	140	1313	1050	875	750	657	584	525	477	438	404	375	350	328	4.43
	125	1146	917	764	655	573	509	458	417	382	353	327	306	286	4.20
914×419	388		§3958	3436	2945	2577	2290	2061	1874	1718	1586	1472	1374	1288	8.73
	343		†3536	3019	2588	2264	2013	1811	1647	1509	1393	1294	1208	1132	8.39
914×305	289	3594	2875	2396	2054	1797	1597	1438	1307	1198	1106	1027	958	899	5.75
	253	3137	2510	2092	1793	1569	1394	1255	1141	1046	965	896	837	784	5.55
	224	2725	2180	1817	1557	1363	1211	1090	991	908	839	779	727	681	5.32
	201	2379	1903	1586	1360	1190	1057	952	865	793	732	680	634	595	5.07
838×292	226	2635	2108	1757	1506	1318	1171	1054	958	879	811	753	703	659	5.45
	194	2194	1755	1463	1254	1097	975	878	798	731	675	627	585	548	5.14
	176	1945	1556	1297	1111	973	864	778	707	648	598	556	519	486	4.94
762×267	197	2057	1646	1371	1176	1029	914	823	748	686	633	588	549	514	5.01
	173	1777	1422	1185	1015	889	790	711	646	592	547	508	474	444	4.77
	147	1479	1184	986	845	740	658	592	538	493	455	423	395	370	4.52
686×254	170	1621	1297	1080	926	810	720	648	589	540	499	463	432	405	4.88
	152	1443	1154	962	824	721	641	577	525	481	444	412	385	361	4.73
	140	1316	1053	877	752	658	585	526	479	439	405	376	351	329	4.60
	125	1149	919	766	656	574	511	459	418	383	353	328	306	287	4.41
610×305	238	2164	1732	1443	1237	1082	962	866	787	721	666	618	577	541	7.03
	179	1621	1297	1080	926	810	720	648	589	540	499	463	432	405	6.40
	149	1350	1080	900	771	675	600	540	491	450	415	386	360	337	6.10
610×229	140	1197	957	798	684	598	532	479	435	399	368	342	319	299	4.48
	125	1063	851	709	608	532	473	425	387	354	327	304	284	266	4.33
	113	950	760	633	543	475	422	380	345	317	292	271	253	238	4.18
	101	830	664	553	474	415	369	332	302	277	255	237	221	207	4.01
533×210	122	924	739	616	528	462	411	369	336	308	284	264	246	231	4.24
	109	816	653	544	467	408	363	327	297	272	251	233	218	204	4.07
	101	758	607	506	433	379	337	303	276	253	233	217	202	190	3.99
	92	685	548	457	391	343	304	274	249	228	211	196	183	171	3.88
	82	593	475	396	339	297	264	237	216	198	183	170	158	148	3.70

Values in the shaded area relate to Universal Beams with tapered flanges.

The loads listed are based on a bending stress of 165 N/mm^2 and assume adequate lateral support. Without such support the span must not exceed L_c unless the compressive stress is reduced in accordance with clause 19a of BS 449.

Loads printed in bold type may cause overloading of the un-stiffened web, the capacity of which should be checked.

Table 9.1 (Continued)

Serial size	Mass per metre	Safe distributed loads in kilonewtons for spans in metres and deflection coefficients													Critical span L$_c$
		2.00	2.50	3.00	3.50	4.00	4.50	5.00	5.50	6.00	7.00	8.00	9.00	10.00	
mm	kg	112.0	71.68	49.78	36.57	28.00	22.12	17.92	14.81	12.44	9.143	7.000	5.531	4.480	m
457 × 191	98	†1066	1033	861	738	645	574	516	469	430	369	323	287	258	4.00
	89	†983	935	779	668	584	519	467	425	389	334	292	260	234	3.86
	82	†911	851	709	608	532	473	426	387	355	304	266	236	213	3.74
	74	†832	771	643	551	482	429	386	351	321	276	241	214	193	3.65
	67	†771	684	570	489	428	380	342	311	285	244	214	190	171	3.53
457 × 152	82	†995	822	685	587	514	457	411	374	343	294	257	228	206	3.03
	74	†913	742	619	530	464	412	371	337	309	265	232	206	186	2.92
	67	825	660	550	471	413	367	330	300	275	236	206	183	165	2.81
	60	†728	591	493	422	370	329	296	269	246	211	185	164	148	2.78
	52	626	501	418	358	313	278	251	228	209	179	157	139	125	2.62
406 × 178	74	†801	699	583	499	437	388	350	318	291	250	218	194	175	3.65
	67	†721	627	523	448	392	348	314	285	261	224	196	174	157	3.54
	60	†634	559	466	399	349	310	279	254	233	200	175	155	140	3.45
	54	611	489	407	349	305	271	244	222	204	174	153	136	122	3.28
406 × 140	46	513	411	342	293	257	228	205	187	171	147	128	114	103	2.58
	39	414	331	276	236	207	184	165	150	138	118	103	92	83	2.41
356 × 171	67	†662	567	472	405	354	315	283	258	236	202	177	157	142	3.72
	57	†574	473	394	338	296	263	237	215	197	169	148	131	118	3.50
	51	†519	420	350	300	263	234	210	191	175	150	131	117	105	3.38
	45	453	363	302	259	227	201	181	165	151	130	113	101	91	3.23
356 × 127	39	377	302	252	216	189	168	151	137	126	108	94	84	75	2.33
	33	311	248	207	177	155	138	124	113	104	89	78	69	62	2.18
305 × 165	54	†479	398	331	284	249	221	199	181	166	142	124	110	99	3.69
	46	†412	342	285	244	214	190	171	155	143	122	107	95	86	3.53
	40	370	296	247	212	185	165	148	135	123	106	93	82	74	3.38
305 × 127	48	404	323	269	231	202	180	162	147	135	115	101	90	81	2.59
	42	351	280	234	200	175	156	140	127	117	100	88	78	70	2.45
	37	311	249	207	178	156	138	124	113	104	89	78	69	62	2.37
305 × 102	33	274	219	183	156	137	122	110	100	91	78	68	61	55	1.90
	28	232	185	154	132	116	103	93	84	77	66	58	51	46	1.79
	25	190	152	127	109	95	84	76	69	63	54	47	42	38	1.64
254 × 146	43	333	267	222	191	167	148	133	121	111	95	83			3.41
	37	286	229	191	164	143	127	115	104	95	82	72			3.22
	31	233	186	155	133	117	104	93	85	78	67	58			2.96
254 × 102	28	203	163	135	116	102	90	81	74	68	58	51			2.01
	25	175	140	117	100	88	78	70	64	58	50	37			1.87
	22	149	119	99	85	74	66	60	54	50	43	37			1.75
203 × 133	30	184	147	123	105	92	82	74	67	61	53				3.03
	25	153	122	102	87	77	68	61	56	51					2.80

Loads printed in italic type do not cause overloading of the unstiffened web, and do not cause deflection exceeding span/360.

Loads printed in ordinary type should be checked for deflection.

† Load is based on allowable shear of web and is less than allowable load in bending.

Table 9.2A Universal beams: dimensions and properties (from BCSA Structural Steelwork Handbook)

Serial size	Mass per metre	Depth of section D	Width of section B	Thickness Web t	Thickness Flange T	Root radius r	Depth between fillets d	Area of section
mm	kg	mm	mm	mm	mm	mm	mm	cm²
914×419	388	920.5	420.5	21.5	36.6	24.1	791.4	493.9
	343	911.4	418.5	19.4	32.0	24.1	791.4	436.9
914×305	289	926.6	307.8	19.6	32.0	19.1	819.1	368.5
	253	918.5	305.5	17.3	27.9	19.1	819.1	322.5
	224	910.3	304.1	15.9	23.9	19.1	819.1	284.9
	201	903.0	303.4	15.2	20.2	19.1	819.1	256.1
838×292	226	850.9	293.8	16.1	26.8	17.8	756.5	288.4
	194	840.7	292.4	14.7	21.7	17.8	756.5	246.9
	176	834.9	291.6	14.0	18.8	17.8	756.5	223.8
762×267	197	769.6	268.0	15.6	25.4	16.5	681.0	250.5
	173	762.0	266.7	14.3	21.6	16.5	681.0	220.2
	147	753.9	265.3	12.9	17.5	16.5	681.0	187.8
686×254	170	692.9	255.8	14.5	23.7	15.2	610.5	216.3
	152	687.6	254.5	13.2	21.0	15.2	610.5	193.6
	140	683.5	253.7	12.4	19.0	15.2	610.5	178.4
	125	677.9	253.0	11.7	16.2	15.2	810.5	159.4
914×419	388	920.5	420.5	21.5	36.6	24.1	799.0	494.5
	343	911.4	418.5	19.4	32.0	24.1	799.0	437.5
914×305	289	926.6	307.8	19.6	32.0	19.1	824.4	368.8
	253	918.5	305.5	17.3	27.9	19.1	824.4	322.8
	224	910.3	304.1	15.9	23.9	19.1	824.4	285.3
	201	903.0	303.4	15.2	20.2	19.1	824.4	256.4
838×292	226	850.9	293.8	16.1	26.8	17.8	761.7	288.7
	194	840.7	292.4	14.7	21.7	17.8	761.7	247.2
	176	834.9	291.6	14.0	18.8	17.8	761.7	224.1
762×267	197	769.6	268.0	15.6	25.4	16.5	685.8	250.8
	173	762.0	266.7	14.3	21.6	16.5	685.8	220.5
	147	753.9	265.3	12.9	17.5	16.5	685.7	188.1
686×254	170	692.9	255.8	14.5	23.7	15.2	615.0	216.6
	152	687.6	254.5	13.2	21.0	15.2	615.0	193.8
	140	683.5	253.7	12.4	19.0	15.2	615.0	178.6
	125	677.9	253.0	11.7	16.2	15.2	615.0	159.6
610×305	238	633.0	311.5	18.6	31.4	16.5	537.2	303.8
	179	617.5	307.0	14.1	23.6	16.5	537.2	227.9
	149	609.6	304.8	11.9	19.7	16.5	537.2	190.1
610×229	140	617.0	230.1	13.1	22.1	12.7	547.2	178.4
	125	611.9	229.0	11.9	19.6	12.7	547.2	159.6
	113	607.3	228.2	11.2	17.3	12.7	547.2	144.5
	101	602.2	227.6	10.6	14.8	12.7	547.2	129.2
533×210	122	544.6	211.9	12.8	21.3	12.7	476.5	155.8
	109	539.5	210.7	11.6	18.8	12.7	476.5	138.6
	101	536.7	210.1	10.9	17.4	12.7	476.5	129.3
	92	533.1	209.3	10.2	15.6	12.7	476.5	117.8
	82	528.3	208.7	9.6	13.2	12.7	476.5	104.4

Values in the shaded area relate to Universal Beams with tapered flanges.

Table 9.2A (Continued)

Serial size	Moment of inertia			Radius of gyration		Elastic modulus		Ratio D/T
	Axis x–x		Axis y–y	Axis x–x	Axis y–y	Axis x–x	Axis y–y	
	gross	Net						
mm	cm⁴	cm⁴	cm⁴	cm	cm	cm³	cm³	
914×419	717325	642716	42481	38.1	9.27	15586	2021	25.2
	623866	559325	36251	37.8	9.11	13691	1733	28.5
914×305	503781	437962	14793	37.0	6.34	10874	961.3	29.0
	435796	379081	12512	36.8	6.23	9490	819.2	32.9
	375111	327298	10425	36.3	6.05	8241	685.6	38.1
	324715	284809	8632	35.6	5.81	7192	569.1	44.7
838×292	339130	315154	10661	34.3	6.08	7971	725.9	31.8
	278833	259610	8384	33.6	5.83	6633	573.6	38.7
	245412	228869	7111	33.1	5.64	5879	487.6	44.4
762×267	239464	221138	7699	30.9	5.54	6223	574.6	30.3
	204747	189347	6376	30.5	5.38	5374	478.1	35.3
	168535	156195	5002	30.0	5.16	4471	377.1	43.1
686×254	169843	156120	6225	28.0	5.36	4902	486.8	29.2
	150015	137962	5391	27.8	5.28	4364	423.7	32.7
	135972	125170	4789	27.6	5.18	3979	377.5	36.0
	117700	108590	3992	27.2	5.00	3472	315.5	41.8
914×419	718742	644135	45407	38.13	9.58	15616	2160	25.2
	625282	560743	39150	37.81	9.46	13722	1871	28.5
914×305	504594	438776	15610	36.99	6.51	10891	1014	29.0
	436610	379896	13318	36.78	6.42	9507	871.9	32.9
	375924	328112	11223	36.30	6.27	8259	738.1	38.1
	325529	285623	9427	35.63	6.06	7210	621.4	44.7
838×292	339747	315771	11353	34.30	6.27	7986	772.9	31.8
	279450	260228	9069	33.63	6.06	6648	620.4	38.7
	246029	229487	7792	33.13	5.90	5894	534.4	44.4
762×267	239894	221568	8174	30.93	5.71	6234	610.0	30.3
	205177	189778	6846	30.51	5.57	5385	513.4	35.3
	168966	156626	5468	29.97	5.39	4483	412.3	43.0
686×254	170147	156424	6621	28.03	5.53	4911	517.7	29.3
	150319	138266	5782	27.85	5.46	4372	454.5	32.7
	136276	125474	5179	27.62	5.38	3988	408.2	36.0
	118003	108894	4379	27.19	5.24	3481	346.1	41.9
610×305	207571	178026	15838	26.14	7.22	6559	1017	20.2
	151631	129960	11412	25.79	7.08	4911	743.3	26.1
	124660	106842	9300	25.61	6.99	4090	610.3	31.0
610×229	111844	101652	4512	25.04	5.03	3626	392.1	27.9
	98579	89634	3933	24.86	4.96	3222	343.5	31.2
	87431	79590	3439	24.60	4.88	2879	301.4	35.1
	75720	69087	2912	24.21	4.75	2515	255.9	40.7
533×210	76207	68609	3393	22.12	4.67	2799	320.2	25.5
	66739	60111	2937	21.94	4.60	2474	278.8	28.7
	61659	55557	2694	21.84	4.56	2298	256.5	30.8
	55353	49912	2392	21.68	4.51	2076	228.6	34.1
	47491	42934	2005	21.32	4.38	1798	192.2	40.0

Values in the shaded area relate to Universal Beams with tapered flanges.

In calculating the net moment of inertia, each flange of 300 mm or greater width is reduced by two holes, and each flange less than 300 mm wide by one hole.

Table 9.2B Universal beams: dimensions and properties (from BCSA Structural Steelwork Handbook)

Serial size	Mass per metre	Depth of section D	Width of section B	Thickness		Root radius r	Depth between fillets d	Area of section
				Web t	Flang T			
mm	kg	mm	mm	mm	mm	mm	mm	cm²
457 × 191	98	467.4	192.8	11.4	19.6	10.2	407.9	125.3
	89	463.6	192.0	10.6	17.7	10.2	407.9	113.9
	82	460.2	191.3	9.9	16.0	10.2	407.9	104.5
	74	457.2	190.5	9.1	14.5	10.2	407.9	95.0
	67	453.6	189.9	8.5	12.7	10.2	407.9	85.4
457 × 152	82	465.1	153.5	10.7	18.9	10.2	406.9	104.5
	74	461.3	152.7	9.9	17.0	10.2	406.9	95.0
	67	457.2	151.9	9.1	15.0	10.2	406.9	85.4
	60	454.7	152.9	8.0	13.3	10.2	407.7	75.9
	52	449.8	152.4	7.6	10.9	10.2	407.7	66.5
406 × 178	74	412.8	179.7	9.7	16.0	10.2	360.5	95.0
	67	409.4	178.8	8.8	14.3	10.2	360.5	85.5
	60	406.4	177.8	7.8	12.8	10.2	360.5	76.0
	54	402.6	177.6	7.6	10.9	10.2	360.5	68.4
406 × 140	46	402.3	142.4	6.9	11.2	10.2	359.6	59.0
	39	397.3	141.8	6.3	8.6	10.2	359.6	49.4
356 × 171	67	364.0	173.2	9.1	15.7	10.2	312.2	85.4
	57	358.6	172.1	8.0	13.0	10.2	312.2	72.2
	51	355.6	171.5	7.3	11.5	10.2	312.2	64.6
	45	352.0	171.0	6.9	9.7	10.2	312.2	57.0
356 × 127	39	352.8	126.0	6.5	10.7	10.2	311.1	49.4
	33	348.5	125.4	5.9	8.5	10.2	311.1	41.8
305 × 165	54	310.9	166.8	7.7	13.7	8.9	265.6	68.4
	46	307.1	165.7	6.7	11.8	8.9	265.6	58.9
	40	303.8	165.1	6.1	10.2	8.9	265.6	51.5
305 × 127	48	310.4	125.2	8.9	14.0	8.9	264.6	60.8
	42	306.6	124.3	8.0	12.1	8.9	264.6	53.2
	37	303.8	123.5	7.2	10.7	8.9	264.6	47.5
305 × 102	33	312.7	102.4	6.6	10.8	7.6	275.8	41.8
	28	308.9	101.9	6.1	8.9	7.6	275.8	36.3
	25	304.8	101.6	5.8	6.8	7.6	275.8	31.4
254 × 146	43	259.6	147.3	7.3	12.7	7.6	218.9	55.1
	37	256.0	146.4	6.4	10.9	7.6	218.9	47.5
	31	251.5	146.1	6.1	8.6	7.6	218.9	40.0
254 × 102	28	260.4	102.1	6.4	10.0	7.6	225.0	36.2
	25	257.0	101.9	6.1	8.4	7.6	225.0	32.2
	22	254.0	101.6	5.8	6.8	7.6	225.0	28.4
203 × 133	30	206.8	133.8	6.3	9.6	7.6	172.3	38.0
	25	203.2	133.4	5.8	7.8	7.6	172.3	32.3

Table 9.2B (Continued)

Serial size	Moment of inertia			Radius of gyration		Elastic modulus		Ratio D/T
	Axis x–x		Axis	Axis	Axis	Axis	Axis	
	Gross	Net	y–y	x–x	y–y	x–x	y–y	
mm	cm⁴	cm⁴	cm⁴	cm	cm	cm³	cm³	
457 × 191	45717	40615	2343	19.10	4.33	1956	243.0	23.9
	41021	36456	2086	18.98	4.28	1770	217.4	26.3
	37103	32996	1871	18.84	4.23	1612	195.6	28.8
	33388	29698	1671	18.75	4.19	1461	175.5	31.6
	29401	26190	1452	18.55	4.12	1296	152.9	35.7
457 × 152	36215	32074	1143	18.62	3.31	1557	149.0	24.6
	32435	28744	1012	18.48	3.26	1406	132.5	27.1
	28577	25357	878	18.29	3.21	1250	115.5	30.6
	25464	22611	794	18.31	3.23	1120	103.9	34.2
	21345	19035	645	17.92	3.11	949.0	84.6	41.3
406 × 178	27329	24062	1545	16.96	4.03	1324	172.0	25.9
	24329	21425	1365	16.87	4.00	1188	152.7	28.6
	21508	18934	1199	16.82	3.97	1058	134.8	31.8
	18626	16457	1017	16.50	3.85	925.3	114.5	37.0
406 × 140	15647	13765	539	16.29	3.02	777.8	75.7	36.0
	12452	11017	411	15.88	2.89	626.9	58.0	46.0
356 × 171	19522	17045	1362	15.12	3.99	1073	157.3	23.2
	16077	14053	1109	14.92	3.92	896.5	128.9	27.5
	14156	12384	968	14.80	3.87	796.2	112.9	30.9
	12091	10609	812	14.57	3.78	686.9	95.0	36.2
356 × 127	10087	9213	357	14.29	2.69	571.8	56.6	33.1
	8200	7511	280	14.00	2.59	470.6	44.7	41.0
305 × 165	11710	10134	1061	13.09	3.94	753.3	127.3	22.7
	9948	8609	897	13.00	3.90	647.9	108.3	26.0
	8523	7384	763	12.86	3.85	561.2	92.4	29.9
305 × 127	9504	8643	460	12.50	2.75	612.4	73.5	22.2
	8143	7409	388	12.37	2.70	531.2	62.5	25.4
	7162	6519	337	12.28	2.67	471.5	54.6	28.4
305 × 102	6487	5800	193	12.46	2.15	415.0	37.8	29.0
	5421	4862	157	12.22	2.08	351.0	30.8	34.8
	4387	3962	120	11.82	1.96	287.9	23.6	44.6
254 × 146	6558	5706	677	10.91	3.51	505.3	92.0	20.4
	5556	4834	571	10.82	3.47	434.0	78.1	23.4
	4439	3879	449	10.53	3.35	353.1	61.5	29.1
254 × 102	4008	3569	178	10.52	2.22	307.9	34.9	26.0
	3408	3046	148	10.29	2.14	265.2	29.0	30.8
	2867	2575	120	10.04	2.05	225.7	23.6	37.2
203 × 133	2887	2476	384	8.72	3.18	279.3	57.4	21.5
	2356	2027	310	8.54	3.10	231.9	46.4	26.0

In calculating the net moment of inertia, each flange of 300 mm or greater width is reduced by two holes, and each flange less than 300 mm wide by one hole.

10 Beams of Two Materials

The basis of design of composite beams is explained by reference to flitch beams, where all the material of each constituent is taken into account in the calculations. The reinforced concrete beam is then dealt with, the elastic theory method receiving a fairly detailed treatment. The load factor method is treated in its simplified form, and the limit state design method is given a very brief outline to provide some idea of the current development trends in structural theory.

FLITCH BEAMS

There are two common cases of beams composed of more than one material. The most usual is that of the reinforced concrete beam, where steel and concrete combine with each other in taking stress. The other case is the **flitch beam** which consists of timber and steel acting together, as shown in Fig. 10.1. A timber joist is strengthened by the addition of a steel plate on the top and bottom, the three members being securely bolted together at intervals.

The bolting together ensures that there is no slip between the steel and the timber at the sections A–A and B–B. Thus the steel and timber at these sections alter in length by the same amount. That is to say,

Strain in the steel at A–A = strain in timber at A–A

But, as shown in Chapter 6,

$$E = \frac{\text{stress}}{\text{strain}} \quad \text{and stress} = \text{strain} \times E$$

Stress in steel at A–A = E steel × strain
Stress in timber at A–A = E timber × (same strain)

Thus

A ──┤████████████├── A   Steel

Timber

Bolts at intervals along beam

B ──┤████████████├── B

Figure 10.1

$$\frac{\text{Stress in steel}}{\text{Stress in timber}} = \frac{E \text{ steel}}{E \text{ timber}}$$

This ratio of the moduli of elasticity is called the **modular ratio** of the two materials and is usually denoted by the letter m. Therefore,

$$\text{Stress in steel} = m \times \text{stress in timber}$$

If, for example, E for steel is 210 000 N/mm² and E for timber is 7000 N/mm², then

$$m = \frac{210\ 000\ \text{N/mm}^2}{7000\ \text{N/mm}^2} = 30$$

and, in effect, the steel plate can thus carry the same load as a timber member of equal thickness but m times the width of the steel.

The beam of the two materials can therefore be considered as an *all-timber beam or equivalent timber beam* as shown in Fig. 10.2.

Similarly, an imaginary *equivalent steel beam* could be formed by substituting for the timber a steel plate having the same depth but only $1/m$th the thickness of the timber, as in Fig. 10.3.

It should be noted that, in forming these equivalent beams, the width only has been altered. Any alteration in the vertical dimensions of the timber or steel would affect the value of the strain and therefore only horizontal dimensions may be altered in forming the equivalent sections.

The strength of the *real* beam may now be calculated by determining the strength of the equivalent timber beam or the equivalent steel beam in the usual manner—treating the section as a normal homogeneous one, but making certain that neither of the maximum permissible stresses (timber and steel) is exceeded.

Figure 10.2 (*a*) Composite beam of timber and steel; (*b*) Equivalent beam of timber

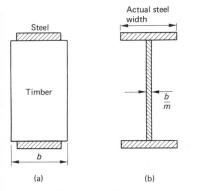

(a) (b)

Figure 10.3 (*a*) Composite beam of timber and steel; (*b*) Equivalent beam of steel

Figure 10.4

Example 10.1

A composite beam consists of a 300 mm × 200 mm timber joist strengthened by the addition of two steel plates 180 mm × 12 mm, as in Fig. 10.4. The safe stress in the timber is 5.5 N/mm², the safe stress in the steel is 165 N/mm² and $m = 30$. Calculate the moment of resistance in N mm.

Solution The equivalent timber beam would be as shown in Fig. 10.5.

I_{xx} of equivalent timber beam is

$$\tfrac{1}{12} \times 200 \times 300^3 + 2(\tfrac{1}{12} \times 5400 \times 12^3 + 5400 \times 12 \times 156^2)$$
$$= (450.0 + 1.6 + 3153.9) \times 10^6 = 3605.5 \times 10^6 \text{ mm}^4$$

Figure 10.5

Figure 10.6

$$Z_{xx} = \frac{3605.5 \times 10^6}{150 + 12} = 22.3 \times 10^6 \text{ mm}^3$$

Stress in timber = 5.5 N/mm² (given)

Therefore

$$M_T = f \times Z = 5.5 \times 22.3 \times 10^6 = 122.4 \times 10^6 \text{ N mm}$$

The equivalent steel beam would be as in Fig. 10.6.

I_{xx} of equivalent steel beam

$$= \tfrac{1}{12} \times 6.7 \times 300^3 + \tfrac{1}{12} \times 180(324^3 - 300^3)$$
$$= (15 + 105.2) \times 10^6 = 120.2 \times 10^6 \text{ mm}^4$$
$$Z_{xx} = 120.2 \times 10^6 / 162 = 0.74 \times 10^6 \text{ mm}^3$$

Stress in steel = 165 N/mm²   (given)

$$\text{or} = 30 \times \text{stress in timber} = 30 \times 5.5$$
$$= 165 \text{ N/mm}^2$$

Therefore

$$M_T = f \times Z = 165 \times 0.74 \times 10^6 = 122.4 \times 10^6 \text{ N mm}$$

Note: In this example, the ratio of the permissible stresses is equal to the modular ratio.

Example 10.2

A composite beam, shown in Fig. 10.7, consists of a 300 mm × 150 mm timber joist strengthened by the addition of two steel plates each 24 mm thick. The stress in the steel must not exceed 150 N/mm²; the stress in the timber must not exceed 8 N/mm²; the modular ratio for the materials is 30. Calculate *a*) the safe moment of resistance of the section in N mm; *b*) the safe uniform load in kN on a span of 4.0 m.

Solution This case has one additional complication which was not present in Example 10.1. The stresses in steel and timber are not to exceed 150 and 8 N/mm², respectively, but it may be that stressing the steel to 150 N/mm² would result in a stress of more than 8 N/mm² in the timber. Alternatively, a stress of 8 N/mm² in the timber may result in causing more than 150 N/mm² in the steel.

The critical case may be reasoned as follows. If the steel at A is stressed to 150 N/mm², then

Stress in the timber at A = 150/30 = 5 N/mm²

From the stress distribution triangles it follows that

Stress in the timber at B = 5 × 150/120 = 6.25 N/mm²

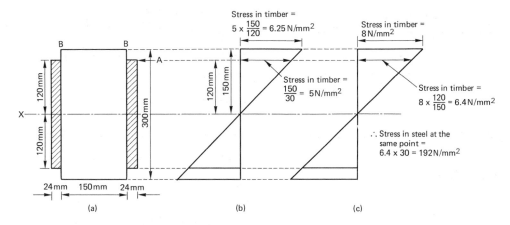

Figure 10.7 (*a*) Composite beam; (*b*) Steel stressed to $150\,\mathrm{N/mm^2}$; (*c*) Timber stressed to $8\,\mathrm{N/mm^2}$

Figure 10.8

Both these stresses (150 and $6.25\,\mathrm{N/mm^2}$) are permissible.

If, on the other hand, the timber at B is stressed to $8\,\mathrm{N/mm^2}$, then (from the triangles of stress distribution)

Stress in the timber at $\mathrm{A} = 8 \times 120/150 = 6.4\,\mathrm{N/mm^2}$

Stress in the steel at $\mathrm{A} = 30 \times 6.4 = 192\,\mathrm{N/mm^2}$.

This stress would be exceeding the safe allowable stress in the steel.

Thus it follows that, in calculating the strength of the composite beam, the stress in the steel may be kept at $150\,\mathrm{N/mm^2}$, but the maximum permissible stress in the equivalent timber section must be reduced to $6.25\,\mathrm{N/mm^2}$.

Fig. 10.8 shows the equivalent timber section.

I_{xx} of equivalent section

$$= 2(\tfrac{1}{12} \times 720 \times 240^3) + \tfrac{1}{12} \times 150 \times 300^3$$
$$= (1658.9 + 337.5) \times 10^6$$
$$= 1996.4 \times 10^6\ \mathrm{mm^4}$$

and

$$Z_{xx} = \frac{1996.4 \times 10^6}{150} = 13.3 \times 10^6\ \mathrm{mm^3}$$

Therefore for *a*)

$$M_T = fZ = 6.25 \times 13.3 = 83.13\ \mathrm{kN\,m}$$

and for *b*)

$$\tfrac{1}{8}W \times 4 = 83.13$$
$$W = 83.13 \times 8/4 = 166.4\ \mathrm{kN}$$

REINFORCED CONCRETE BEAMS

Concrete is a material strong in its resistance to compression, but very weak indeed in tension. A good concrete will safely take a stress upward of 7 N/mm² in compression, but the safe stress in tension is usually limited to no more than $\frac{1}{10}$th of its compressive stress. It will be remembered that in a homogeneous beam the stress distribution is as shown in Fig. 10.9, and in the case of a section symmetrical about the X–X axis, the actual stress in tension equals the actual stress in compression. If such a beam was of ordinary unreinforced concrete, then the stresses in tension and in compression would of necessity have to be limited to avoid over-stressing in the tension face, whilst the compressive fibres would be taking only $\frac{1}{10}$th of their safe allowable stress. The result would be most uneconomical since the compression concrete would be under-stressed.

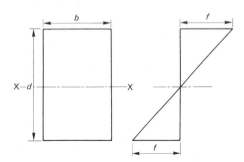

Figure 10.9

In order to increase the safe load carrying capacity of the beam, and allow the compression concrete to use to the full its compressive resistance, steel bars are introduced in the tension zone of the beam to carry the whole of the tensile forces.

This may, at first, seem to be a little unreasonable—for the concrete is capable of carrying some tension, even though its safe tensile stress is very low. In fact it will be seen from Fig. 10.10 that the strain in the steel and in the concrete at A–A are the same. Therefore, if the stress in steel is 140 N/mm², the concrete would be apparently stressed to $1/m$th of this and, since m, the ratio of the elastic moduli of steel and concrete, is usually taken as 15, the tensile stress in concrete would be in excess of 9 N/mm². The concrete would crack through failure in tension at a stress very much lower than this and thus its resistance to tension is disregarded.

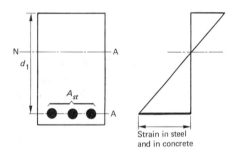

Strain in steel and in concrete

Figure 10.10

BASIS OF DESIGN: 1) ELASTIC THEORY METHOD

In the design of reinforced concrete beams the following assumptions are made:

1 Plane sections through the beam before bending remain plane after bending.

2 The concrete above the neutral axis carries *all* the compression.

3 The tensile steel carries *all* the tension.

4 Concrete is an elastic material, and therefore stress is proportional to strain.

5 The concrete shrinks in setting and thus grips the steel bars firmly so that there is no slip between the steel and the surrounding concrete.

As the concrete below the neutral axis (NA) is ignored from the point of view of its tensile strength, the actual beam shown in Fig. 10.11(a) is, in fact, represented by an area of concrete above the neutral axis carrying compression, and a relatively small area of steel taking tension as in Fig. 10.11(b). Thus the position of the neutral axis will not normally be at $d_1/2$ from the top of the beam, but will vary in position according to the relation between the amount of concrete above the NA and the area of steel.

The distance from the top face of the beam to the NA which indicates the depth of concrete in compression in a beam is denoted by d_n.

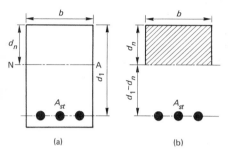

Figure 10.11

Critical Area of Steel

In a beam of a given size, say 200 mm wide with a 300 mm effective depth, the area of steel could be varied—it could be 1000 mm^2 in area, or 2000 mm^2, etc., and each addition of steel would make for a stronger beam, but not necessarily for an economical one.

If, for example, only a very small steel area is included, as in Fig. 10.12, and this small steel content is stressed up to its safe allowable value, then the concrete will probably be very much under-stressed, so that the concrete is not used to its full advantage. From this point of view, an uneconomical beam results.

Figure 10.12

If, on the other hand, an over-generous amount of steel is included, as in Fig. 10.13, then stressing this steel to its safe allowable value will probably result in over-stressing the concrete. Thus, to keep the concrete stress to its safe amount, the stress in the steel will have to be reduced, and again the result is an uneconomical section.

Given that $m = 15$, there will be only one critical area of steel for a given size of beam which will allow the stress in the concrete to reach its permissible value at the same time as the stress in the steel reaches its maximum allowable value. In reinforced concrete beam calculations the designer is nearly always seeking to choose a depth of beam and an area of steel which will allow these stresses to be attained together thus producing what is called a "balanced" or "economical" section. The breadth of the beam is normally chosen from practical considerations.

Figure 10.13

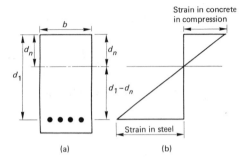

Figure 10.14

If the two given safe stresses are reached together, then the required area of steel will always remain a constant fraction of the beam's area (bd_1) and the factor d_n/d_1 will also have a constant value.

Fig. 10.14(a) shows a normal reinforced concrete (r.c.) beam and (b) the strain diagram for the beam.

From similar triangles it will be seen that

$$\frac{\text{Strain in concrete}}{\text{Strain in steel}} = \frac{d_n}{d_1 - d_n} \tag{1}$$

But stress/strain for any material is E (Young's modulus) for the material.

Therefore   strain = stress/E

Let p_{cb} represent the permissible compressive stress for concrete in bending in N/mm^2 and let p_{st} represent the permissible tensile stress in the steel in N/mm^2.

Thus from equation (1)

$$\frac{d_n}{d_1 - d_n} = \frac{\text{stress in concrete}}{E_c} \div \frac{\text{stress in steel}}{E_s}$$

$$\frac{d_n}{d_1 - d_n} = \frac{p_{cb}}{E_c} \div \frac{p_{st}}{E_s} = \frac{p_{cb}}{E_c} \times \frac{E_s}{p_{st}}$$

But $E_s/E_c = m$, modular ratio, therefore

$$\frac{d_n}{d_1 - d_n} = \frac{m p_{cb}}{p_{st}} \tag{2}$$

Equation (2) shows that d_n and d_1 will always be proportional to each other when m, p_{cb}, and p_{st} are constant.

From equation (2)

$$d_1 - d_n = \frac{d_n p_{st}}{m p_{cb}}$$

$$d_1 = d_n \left(1 + \frac{p_{st}}{m p_{cb}}\right)$$

So

$$d_n = \frac{d_1}{1 + (p_{st}/m p_{cb})} \tag{3}$$

Note: d_n is the distance down to the neutral axis from the compression face of the beam, and is expressed in millimetres.

As d_n is, however, seen to be a constant fraction of d_1 for fixed values of m, p_{cb} and p_{st}, it is frequently useful to know the value of the constant d_n/d_1, and this constant is usually denoted by n, i.e.

$$d_n/d_1 = n$$

As d_n is equal to $\dfrac{d_1}{1+p_{st}/mp_{cb}}$ from equation (3), then

$$n = \frac{d_n}{d_1} = \frac{1}{1+(p_{st}/mp_{cb})} \qquad (4)$$

For the values $m = 15$, $p_{cb} = 7\ \text{N/mm}^2$, $p_{st} = 140\ \text{N/mm}^2$

$$n = \frac{1}{1+\dfrac{140}{15\times 7}} = 0.43$$

and $d_n = 0.43d_1$

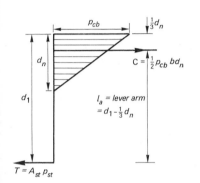

Figure 10.15

Fig. 10.15 shows that the stress in the concrete from the top of the section to the NA follows the normal distribution for ordinary homogeneous sections, and the total compression force $C = p_{cb}bd_n/2$, i.e. an area bd_n of concrete, taking stress at an average rate of $p_{cb}/2\ \text{N/mm}^2$.

The tension, however, will now all be taken by the steel and will be $T = A_{st} \times p_{st}$, i.e. an area of steel A_{st} taking stress at $p_{st}\ \text{N/mm}^2$. These two internal forces T and C are equal and opposite in their direction, and they form an internal couple acting at $d_1 - \tfrac{1}{3}d_n$ apart, where $d_1 - \tfrac{1}{3}d_n$ is the lever arm, denoted by l_a.

This internal couple or moment of resistance may thus be written as follows:

i) in terms of concrete $\tfrac{1}{2}p_{cb}bd_n(d_1 - \tfrac{1}{3}d_n) = M_{rc}$ (5)

$\left.\right\}M_{max}$

ii) in terms of steel $A_{st} \times p_{st} \times (d_1 - \tfrac{1}{3}d_n) = M_{rt}$ (6)

But just as d_n/d_1 is a "constant n," similarly $(d_1 - \tfrac{1}{3}d_n)/d_1$ is also a constant, say, j and so the lever arm l_a may be expressed as jd_1.

For the values $m = 15$, $p_{cb} = 7\ \text{N/mm}^2$, $p_{st} = 140\ \text{N/mm}^2$,

$$j = \frac{d_1 - \tfrac{1}{3}d_n}{d_1} = \frac{1 - \tfrac{1}{3}0.43}{1} = 0.86$$

and $l_a = 0.86d_1$

Substituting the constants n and j in (5),

$$M_{rc} = \tfrac{1}{2}p_{cb}bd_n(d_1 - \tfrac{1}{3}dn) = \tfrac{1}{2}p_{cb} \times b \times nd_1 \times jd_1$$
$$= \tfrac{1}{2}p_{cb} \times n \times jbd_1^2$$

As $\tfrac{1}{2}p_{cb}$, n and j are all constants, then $\tfrac{1}{2}p_{cb} \times n \times j$ has a constant value which may be called R, and therefore

$$M_{rc} = Rbd_1^2 \qquad (7)$$

Using the values $m = 15$, $p_{cb} = 7\ \text{N/mm}^2$, $p_{st} = 140\ \text{N/mm}^2$

$$R = \tfrac{1}{2}p_{cb} \times n \times j = \tfrac{1}{2} \times 7 \times 0.43 \times 0.86 = 1.29$$

and

$$M_{rc} = 1.29bd_1^2$$

Hence

$$d_1^2 = \frac{M}{1.29b} \quad \text{and} \quad d_1 = \sqrt{\frac{M}{1.29b}} \tag{8}$$

As $M_{rt} = A_{st} \times p_{st} \times jd_1$ then the required area of steel is

$$A_{st} = \frac{M}{p_{st} \times jd_1} = \frac{M}{p_{st}l_a} \tag{9}$$

Note: As the area of steel is a constant fraction of the area $b \times d_1$ (i.e. $A_{st} = bd_1 \times$ a constant), this relationship may be specified as the percentage area of steel, p, where

$$p\% = \frac{A_{st}}{bd_1} \times 100 \quad \text{or} \quad A_{st} = bd_1 \times \frac{p}{100}$$

From equation (9)

$$A_{st} = \frac{M}{p_{st}jd_1}$$

but

$$M = Rbd_1^2 = \tfrac{1}{2}p_{cb} \times n \times j \times bd_1^2$$

therefore

$$A_{st} = \frac{p_{cb} \times n \times j \times bd_1^2}{2 \times p_{st} \times jd_1} = \frac{p_{cb} \times n \times bd_1}{2 \times p_{st}}$$

$$\frac{A_{st}}{bd_1} = \frac{p_{cb}n}{2p_{st}}$$

So

$$p = \text{percentage of steel} = \frac{A_{st}}{bd_1} \times 100 = \frac{p_{cb} \times n \times 100}{2p_{st}}$$

$$p\% = \frac{50 \times p_{cb} \times n}{p_{st}} \tag{10}$$

Using the values $p_{cb} = 7 \text{ N/mm}^2$, $p_{st} = 140 \text{ N/mm}^2$, $m = 15$,

$$p = \frac{50 \times 7 \times 0.43}{140} = 1.07 \text{ per cent}$$

Hence

$$A_{st} = bd_1 \times (1.07/100) = 0.011bd_1$$

It must be remembered, however, that there is a wide range of concrete strengths, hence it is not advisable to memorise the numerical values of these design constants.

They may be derived, quite easily, from

$$n = \frac{1}{1 + (p_{st}/mp_{cb})} \qquad j = 1 - \tfrac{1}{3}n$$

$$R = \tfrac{1}{2}p_{cb} \times n \times j \qquad p\% = \frac{50 \times p_{cb} \times n}{p_{st}}$$

From all the above it will be seen that in designing simple reinforced concrete beams the procedure is as follows:

1 Calculate maximum bending moment, M_{max}, in N mm.

2 Choose a suitable breadth of beam, b.

3 Calculate effective depth to centre of steel, $d_1 = \sqrt{\dfrac{M_{max}}{R \times b}}$

4 Determine area of steel required from

$$A_{st} = \frac{M_{max}}{p_{st} \times j \times d_1} \quad \text{or} \quad A_{st} = \frac{pbd_1}{100}$$

Example 10.3

Design a simple reinforced concrete beam 150 mm wide to withstand a maximum bending moment of 20×10^6 N mm using the following permissible stresses: $p_{cb} = 7$ N/mm², $p_{st} = 140$ N/mm² and $m = 15$.

Solution
a) Solution by design constants:

$$n = \frac{1}{1 + (p_{st}/mp_{cb})} = \frac{7}{7 + 140/15} = 0.43$$

$$j = 1 - \tfrac{1}{3}n = 1 - \tfrac{1}{3}0.43 = 0.86$$

$$R = \tfrac{1}{2}p_{cb} \times n \times j = \tfrac{1}{2} \times 7 \times 0.43 \times 0.86 = 1.29$$

$$p\% = \frac{50 \times p_{cb} \times n_1}{p_{st}} = 50 \times 0.43 \times 7/140 = 1.07$$

$$d_1 = \sqrt{\frac{20\,000\,000}{1.29 \times 150}} = \sqrt{103\,352} = 321 \text{ mm}$$

$$A_{st} = 0.0107 \times 150 \times 321 = 516 \text{ mm}^2$$

b) Alternative solution:
From equation (3)

$$d_n = \frac{d_1}{1 + \dfrac{140}{15 \times 7}} = \tfrac{3}{7}d_1 = 0.43d_1$$

From equation (5) and substituting $0.43d_1$ for d_n

$$M_{rc} = \tfrac{1}{2}p_{cb} \times b \times 0.43d_1 \times (d_1 - \tfrac{1}{3}0.43d_1)$$

Figure 10.16

$$20 \times 10^6 = \tfrac{1}{2} \times 7 \times 150 \times 0.43 \times 0.86 \times d_1^2$$

$$d_1 = \sqrt{\frac{20\,000\,000 \times 2}{7 \times 150 \times 0.43 \times 0.86}} = 321 \text{ mm}$$

From equation (6)

$$A_{st} = \frac{M}{p_{st} \times 0.86 \times d_1} = \frac{20\,000\,000}{140 \times 0.86 \times 321} = 516 \text{ mm}^2$$

Use five 12 mm diameter bars (area $= 5 \times 113.1 = 566$ mm^2). The overall depth d is determined as follows.

The effective depth d_1 is measured from the top face to the centre of the steel bars. The bars themselves need an effective cover of at least 25 mm, so to determine the overall depth d it is necessary to add (half the bar diameter $+25$ mm) to the effective depth d_1 (see Fig. 10.16).

In this case overall depth $d = 322 + 6 + 25 = 353$ mm. Say 150 mm $\times$ 355 mm overall with five 12 mm diameter bars.

CP114:1969 distinguishes two types of concrete mixes—the nominal and the designed concrete mix. The permissible stresses in compression due to bending, p_{cb}, for nominal concrete mixes are as follows:

$$1:1\ :2 \quad 10.0 \text{ N/mm}^2$$
$$1:1\tfrac{1}{2}:3 \quad 8.5 \text{ N/mm}^2$$
$$1:2\ :4 \quad 7.0 \text{ N/mm}^2$$

Note that these stresses assume that the concrete has cube crushing strengths as specified in the Code. When difficulty is found in reaching the requisite cube strengths, figures 25 per cent lower than those given may be accepted, provided that the permissible stresses are reduced in the same ratio. Note also that a "nominal 1:2:4 mix" means 50 kg of cement to 0.07 m^3 of fine aggregate and 0.14 m^3 of coarse aggregate. The other mixes are in similar proportions.

For designed concrete mixes higher p_{cb} values may be used.

The permissible tensile stress p_{st} for hot rolled mild steel bars is 140 N/mm^2. Steels with certain guaranteed yield points, and special steels, are permitted higher stresses.

Example 10.4

A simply supported reinforced concrete beam is to span 5 m carrying a total uniform load of 60 kN inclusive of self weight, and a point load of 90 kN from a secondary beam as shown in Fig. 10.17. The beam is to be 200 mm wide. Choose a suitable overall depth and area of tensile steel reinforcement for the maximum bending moment. Assume: $p_{cb} = 8.5$ N/mm^2, $p_{st} = 140$ N/mm^2, $m = 15$.

Figure 10.17

Solution

$$R_R = \tfrac{1}{2} \times 60 + \tfrac{1}{5} \times 90 \times 1 = 30 + 18 = 48 \text{ kN}$$
$$R_L = 60 + 90 - 48 = 102 \text{ kN}$$

M_{max} occurs at a distance

$$x = \frac{(102 - 90) \times 5}{60} = 1 \text{ m}$$

from R_R, i.e. at the point load.

$$M_{max} = 102 \times 1 - \tfrac{1}{5} \times 60 \times \tfrac{1}{2} = 96 \times 10^6 \text{ N mm}$$

$$d_n = \frac{8.5}{8.5 + (140/15)} = 0.48 d_1 \quad \text{and} \quad l_a = 0.84 d_1$$

$$d_1 = \sqrt{\frac{2 \times 96\,000\,000}{8.5 \times 200 \times 0.48 \times 0.84}} = 530 \text{ mm}$$

$$A_{st} = \frac{96\,000\,000}{140 \times 0.84 \times 530} = 1540 \text{ mm}^2$$

Use five 20 mm diameter bars (1570 mm²).
Then $d = 530 + \tfrac{1}{2} \times 20 + 25 = 565$ mm.

Example 10.5

A simply supported beam is to span 3.6 m carrying a uniform load of 60 kN inclusive of self weight. The beam is to be 150 mm wide, and the stresses in steel and concrete respectively are to be 140 N/mm² and 10 N/mm² ($m = 15$).

Determine the constants n, j, R and $p\%$, and choose a suitable effective depth, overall depth and area of steel for the beam.

Solution

$$n = \frac{1}{1 + (p_{st}/mp_{cb})} = \frac{1}{1 + (140/15 \times 10)} = 0.517$$

$$j = 1 - \tfrac{1}{3}n = 1 - \tfrac{1}{3}0.517 = 0.828$$

$$R = \tfrac{1}{2}p_{cb} \times n \times j = \tfrac{1}{2}10 \times 0.517 \times 0.828 = 2.140$$

$$p\% = \frac{50 \times p_{cb} \times n}{p_{st}} = \frac{50 \times 10 \times 0.517}{140} = 1.846$$

$$M_{max} = \tfrac{1}{8}Wl = \tfrac{1}{8} \times 60 \times 3.6 \times 10^6 = 27 \times 10^6 \text{ N mm}$$

$$d_1 = \sqrt{\frac{27\,000\,000}{2.140 \times 150}} = 290 \text{ mm}$$

$$A_{st} = 0.01846 \times 150 \times 290 = 803 \text{ mm}^2$$

Use four 16 mm diameter bars (804 mm²).
Then $d = 290 + \tfrac{1}{2}16 + 25 = 323$, say 325 mm.

BASIS OF DESIGN: 2) LOAD FACTOR METHOD

The introduction of this method alongside the elastic theory method to CP114 in 1957 was the result of tests to destruction. They have shown that, at failure, the compressive stresses adjust themselves to give a compressive resistance greater than that obtained by the elastic theory method.

The load factor method, therefore, does not use the modular ratio nor does it assume a proportionality between stress and strain in concrete. It requires, as mentioned in Chapter 6, the knowledge of plastic behaviour, which is outside the scope of this textbook. Clause 306 of CP114, however, gives simplified formulae for rectangular beams and slabs. They are based on the assumption that the compressive stress in the concrete is two-thirds of the permissible compressive stress in the concrete in bending (i.e. $\frac{2}{3}p_{cb}$). The stress is considered to be uniform over a depth d_n, not exceeding one-half of the effective depth $\frac{1}{2}d_1$.

Fig. 10.18 shows the stress distribution diagram based on the above assumptions.

Figure 10.18

Total compressive force $C = \frac{2}{3}p_{cb} \times b \times d_n$

Total tensile force $T = p_{st} \times A_{st}$

For equilibrium $C = T$ and therefore

$$d_n = \frac{3}{2} \times \frac{p_{st} \times A_{st}}{p_{cb} \times b}$$

Since the compressive stress is uniform

Lever arm $l_a = d_1 - \frac{1}{2}d_n$

$$= d_1 - \frac{3 \times p_{st} \times A_{st}}{4 \times p_{cb} \times b}$$

but, for an "economical" or balanced section, the Code recommends $d_n = \frac{1}{2}d_1$

$$l_a = d_1 - \frac{1}{2} \times \frac{1}{2}d_1 = \frac{3}{4}d_1$$

Hence, the moment of resistance, based on concrete in compression, is

$$M_{rc} = \frac{2}{3}p_{cb} \times b \times \frac{1}{2}d_1 \times \frac{3}{4}d_1$$
$$= \frac{1}{4}p_{cb} \times b \times d_1^2$$

and the moment of resistance, based on tensile reinforcement, is

$$M_{rt} = A_{st} \times p_{st} \times l_a$$

Example 10.6

Consider the beam in Example 10.5 and, using the same concrete mix, design the beam by the load factor method.

Solution

$$M_{max} = 27 \times 10^6 \text{ N mm}$$
$$= M_{rc} = \tfrac{1}{4} 10 \times 150 \times d_1^2$$
$$d_1 = \sqrt{\frac{27\,000\,000 \times 4}{10 \times 150}} = 268 \text{ mm}$$

and

$$A_{st} = \frac{27\,000\,000 \times 4}{140 \times 3 \times 268} = 960 \text{ mm}^2$$

It may be correctly deduced from the above that the load factor method of design results in shallower beams with more reinforcement.

The load factor method gained fairly general acceptance in relatively short time, so that, when in 1972 the new Code of Practice for the Structural Use of Concrete (CP110) appeared, it did not contain any explicit reference to the modular ratio or elastic theory.

This Code introduces yet another approach to structural design in, what it calls, "strict conformity with the theory of limit states". It is published in three parts—the first dealing with Design, Materials and Workmanship, the remaining two containing Design Charts prepared in accordance with the assumptions laid down in Part 1.

BASIS OF DESIGN: 3) LIMIT STATE DESIGN

A limit state may be defined as that state of a structure when it becomes unfit for the use for which it was designed.

To satisfy the object of this method of design, all relevant limit states should be considered in order to ensure an adequate degree of safety and serviceability. The Code, therefore, distinguishes between the **limit state of collapse,** which concerns the safety of the structure and is referred to as the ultimate limit state, and the **limit states of serviceability,** which deal with deflection, cracking of concrete, vibration, etc.

Whereas the other two design methods relied on a single factor of safety (or load factor), CP110 introduces two partial safety factors, one applied to the strength of the materials and the other to loads. It, therefore, enables the designer to vary the degree of risk by choosing different partial safety factors. In this way the limit state design ensures an acceptable probability that the limit states will not be reached and so is able to provide a safe and serviceable structure economically.

Consider the design of a reinforced concrete beam for the ultimate limit state.

1 Loads The design load is obtained by multiplying the characteristic load by an appropriate partial safety factor, γ_f.

The characteristic loads (G_k = dead load, Q_k = imposed load, and W_k = wind load) are based on values given in CP3, Chapter V Parts 1 and 2.

The partial safety factors γ_f depend on the combination of loads. For example, for dead + live loads, the design load would be

$$(1.4 \times G_k + 1.6 \times Q_k)$$

and for dead + live + wind loads, the design load would be

$$1.2 \times (G_k + Q_k + W_k)$$

2 Strength of materials The design strength is obtained by dividing the characteristic strength f_k, by the appropriate partial safety factor γ_m.

Characteristic strength is defined as that value of the cube strength of concrete (f_{cu}) or the yield (or proof) stress of reinforcement (f_y) below which not more than 5% of the test results will fall.

The partial safety factors γ_m for ultimate limit state are 1.5 for concrete and 1.15 for the reinforcement.

3 Ultimate moment of resistance The Code gives a simplified version of calculations based on assumptions similar to the load factor method, i.e. rectangular stress block with the stress value $0.4 \times f_{cu}$.

Therefore

$$M_{uc} = 0.4 f_{cu} \times b \times \tfrac{1}{2}d \times \tfrac{3}{4}d$$
$$= 0.15 f_{cu} \times b \times d^2$$

and

$$M_{ut} = 0.87 f_y \times A_s \times z$$

where M_{uc} = ultimate resistance moment (compression)
M_{ut} = ultimate resistance moment (tension)
b = width of beam section
d = effective depth to tension reinforcement
A_s = area of tension reinforcement
z = lever arm (in this case $\tfrac{3}{4}d$).

Example 10.7

Take the beam in Example 10.5 and adapt the data to limit state design method for the ultimate limit state.

Solution Assume the 60 kN load to consist of 60% dead and 40% live loads and

$$f_y = 250 \text{ N/mm}^2 \qquad f_{cu} = 30 \text{ N/mm}^2$$

Therefore,

$$\text{Design load} = (1.4 \times 36 + 1.6 \times 24) \text{ kN}$$

and

$$M_{max} = \tfrac{1}{8} \times 88.8 \times 3.6 \times 10^6 = 40 \times 10^6 \text{ N mm}$$

Hence

$$M_{uc} = 0.15 \times 30 \times 150 \times d^2 = 40 \times 10^6 \text{ N mm}$$

$$d = \sqrt{\frac{40\,000\,000}{0.15 \times 30 \times 150}} = 243 \text{ mm}$$

and

$$M_{ut} = 0.87 \times 250 \times A_s \times \tfrac{3}{4} \times 243 = M_{max}$$

$$A_s = \frac{40\,000\,000}{0.87 \times 250 \times \tfrac{3}{4} \times 243} = 1009 \text{ mm}^2$$

It must be emphasised that the above exposition of the limit state design method is very much simplified and it is essential to consult textbooks on concrete design for more detailed discussion of both principles and their application.

SUMMARY

Timber and Steel Composite (Flitch) Beams

Replace the steel by its equivalent area of timber by multiplying the width of the steel (parallel to the axis of bending) by the modular ratio E_s/E_t. Calculate the value of Z for the equivalent timber section. Determine the maximum permissible stress f in the extreme fibres of the beam such that neither the maximum permissible steel stress nor the maximum permissible timber stress is exceeded. Then

$$M_r = fZ$$

Reinforced Concrete Beams

a) *Elastic Theory Method*

i) $d_n = \dfrac{1}{1 + p_{st}/mp_{cb}} \times d_1$

ii) $M_{rc} = \frac{1}{2}p_{cb} \times b \times d_n \times l_a$

iii) $M_{rt} = p_{st} \times A_{st} \times l_a$

where $l_a = d_1 - \frac{1}{3}d_n$

b) *Load Factor Method*

$M_{rc} = \frac{1}{4}p_{cb} \times b \times d_1^2$

$M_{rt} = p_{st} \times A_{st} \times \frac{3}{4}d_1$

c) *Limit State Design Method*

$M_{uc} = 0.15 f_{cu} \times b \times d^2$

$M_{ut} = 0.87 f_y \times A_s \times z$

where $z = \frac{3}{4}d$.

Figure 10.19

Figure 10.20

EXERCISE 10

1 A composite beam is formed using a $400 \text{ mm} \times 180 \text{ mm}$ timber beam with a $300 \text{ mm} \times 12 \text{ mm}$ steel plate securely fixed to each side as shown in Fig. 10.19. The maximum stresses in the steel and timber respectively must not exceed 140 and 8 N/mm^2, and the modular ratio is 20.

a) What will be the actual stresses used for i) the steel, and ii) the timber?

b) What is the safe moment of resistance in N mm for the beam section?

2 A timber flitch beam is composed of two $300 \text{ mm} \times 150 \text{ mm}$ timber beams and one $250 \text{ mm} \times 20 \text{ mm}$ steel plate placed between the timbers so that, when properly bolted together, the centre lines of all three members coincide. Calculate the maximum safe uniformly distributed load in kilonewtons that this beam could carry over a span of 4.5 m if the stress in the steel is not to exceed 125 N/mm^2 and that in the timber 7 N/mm^2, and given that the modular ratio $E_s/E_t = 20$.

3 A timber beam $150 \text{ mm} \times 300 \text{ mm}$ deep has two steel plates, each $125 \text{ mm} \times 12 \text{ mm}$ bolted to it as shown in Fig. 10.20. Assuming the safe steel stress is 140 N/mm^2, the safe timber stress is 7 N/mm^2, E for steel is $210\,000 \text{ N/mm}^2$ and E for timber is 8400 N/mm^2, calculate the moment of resistance of the beam. (Ignore bolt holes.)

Figure 10.21

Figure 10.22

Figure 10.23

Figure 10.24

4 A timber beam in an existing building is 200 mm wide and 380 mm deep and is simply supported at the ends of a 6 m span. *a*) Calculate the maximum safe uniformly distributed load for the timber alone if the bending stress must not exceed 7 N/mm². *b*) It is proposed to strengthen the beam to enable it to carry a uniformly distributed load of 150 kN by bolting two steel plates to the beam as indicated in Fig. 10.21. Calculate the required thickness *t* of the plates if the maximum permissible stress for the steel is 140 N/mm² and the modular ratio is 24.

5 A short concrete beam is to be constructed without any steel reinforcement to span 2.8 m, carrying a total inclusive uniform load of 20 kN. If the concrete has a safe tensile stress of only 0.6 N/mm², state what depth would be needed for a suitable beam 200 mm wide.

6 Design a reinforced concrete section for the loading conditions as in Question 5 if the beam remains 200 mm wide, using the following stresses: $p_{cb} = 10$ N/mm², $p_{st} = 140$ N/mm², $m = 15$.

7 A simply supported reinforced concrete beam, 240 mm wide, carries inclusive loads as shown in Fig. 10.22. Determine *a*) the effective depth d_1 in mm, *b*) the required steel area if $p_{cb} = 7$ N/mm², $p_{st} = 140$ N/mm², $m = 15$. The weight of the beam may be assumed to be included in the given loads.

8 Referring to Example 9.5 design a reinforced concrete beam as an alternative to the 400 mm × 150 mm timber beam. Assume that the floor load is 7 kN/m² and allow 12 kN for the weight of the reinforced concrete beam. Take the breadth of the beam as 250 mm and $p_{cb} = 7$ N/mm², $p_{st} = 140$ N/mm², $m = 15$.

9 Referring to Example 9.7 for loading conditions but substituting 18 kN for the weight of the beam, design a reinforced concrete beam 225 mm wide. When the design is complete, check the assumed weight of beam, taking the weight of reinforced concrete as 24 kN/m³. Use $1 : 1\frac{1}{2} : 3$ mix of concrete.

10 A small floor is to be supported as shown in Fig. 10.23. The total floor load is 5 kN/m², and 12 kN can be assumed as the weight of the beam. Design the beam assuming a breadth of 250 mm and $1 : 2 : 4$ mix concrete.

11 Referring to Exercise 9.15, but taking 8 kN as the weight of each beam instead of the value given, design reinforced concrete beams assuming a breadth of 225 mm and $1 : 1 : 2$ mix concrete.

12 Referring to Exercise 9.16 design beams A and B in reinforced concrete ($1 : 2 : 4$ mix). Take the weight of beam B as 1.8 kN and the weight of beam A as 16.0 kN (in addition to the floor load given).

13 A reinforced concrete beam 300 mm wide simply supported on a span of 6 m carries a triangular load of 80 kN in addition to its own weight, which may be assumed to be 20 kN (Fig. 10.24). Design the beam in $1 : 1\frac{1}{2} : 3$ mix concrete.

11 Deflection of Beams

So far in this book the beam has been considered from the point of view of its safety and strength, in its resistance to bending. This chapter investigates the deformation of beams as the direct effect of that bending tendency, which affects their serviceability and stability, and does so in terms of their deflection.

A beam may be strong enough to resist safely the bending moments due to the applied loading and yet not be suitable because its deflection is too great. Excessive deflection might not only impair the strength and stability of the structure but also give rise to minor troubles such as cracking of plaster ceilings, partitions and other finishes, as well as adversely affecting the functional needs and aesthetic requirements or simply being unsightly.

BS449 for steel, and the relevant Codes of Practice for other materials, stipulate that the deflection of a beam shall be restricted within limits appropriate to the type of structure. In the case of structural steelwork the maximum deflection due to loads other than dead loads must not exceed 1/360 of the span (Cl.15, BS449: Part 2: 1969). For timber beams, on the other hand, the figure is 0.003 of the span when the supporting member is fully loaded (Cl.3.13.2, CP112: Part 2: 1971). In reinforced concrete the deflection is generally governed by the span/depth ratio (Cl.309, CP114: Part 2: 1969; Cl.3.3.8, CP110: Part 1: 1972).

FACTORS AFFECTING DEFLECTION

For many beams in most types of buildings, e.g. flats, offices, warehouses, it will usually be found that, if the beams are made big enough to resist the bending stresses, the deflections will not exceed the permitted values. In beams of long spans, however, it may be necessary to calculate deflections to ensure that they are not excessive. The derivation of formulae for calculating deflections usually involves the calculus. In this chapter, therefore, only a general treatment will be attempted and deflection formulae for a few common cases of beam loadings will be given without proof. General methods of calculation of deflections are given in standard books on theory of structures or strength of materials.

Figure 11.1

Load

AB (Fig. 11.1) represents a beam of span l metres supported simply at its ends and carrying a point load of W kN at mid-span. Let us assume that the deflection due to the load is 5 mm. It is obvious that, if the load is increased, the deflection will increase. It can be proved that the deflection is directly proportional to the load, i.e. a load of $2W$ will cause a deflection of 10 mm, $3W$ will produce a deflection of 15 mm and so on. W must therefore be a term in any formula for calculating deflection.

Span

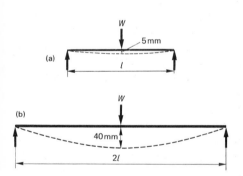

Figure 11.2

In Fig. 11.2(a) and (b) the loads W are equal and the weights of the beams, which are assumed to be equal in cross-section, are ignored for purposes of this discussion. The span of beam b is twice that of beam a. It is obvious that the deflection of beam b will be greater than that of beam a, but the interesting fact (which can be demonstrated experimentally or proved by mathematics) is that, instead of the deflection of b being twice that of a, it is 8 times (e.g. 40 mm in this example). If the span of beam b were $3l$, its deflection would be 27 times that of beam a. In other words, the deflection of a beam is proportional to the cube of the span, therefore l^3 is a term in the deflection formula.

Size and Shape of Beam

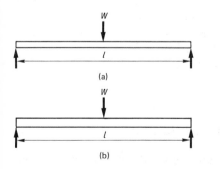

Figure 11.3 (a) Moment of inertia of beam = 1 unit; (b) Moment of inertia of beam = 2 units

Fig. 11.3(a) and (b) represents two beams (their weights being ignored) of equal spans and loading but the moment of inertia of beam b is twice that of beam a. Obviously, the greater the size of the beam, the less the deflection (other conditions being equal). It can be proved that the deflection is inversely proportional to the moment of inertia, e.g. the deflection of beam b will be one half of that of beam a. Moment of inertia I is therefore a term in the denominator of the deflection formula. (It may be noted that, since the moment of inertia of a rectangular cross-section beam is $bd^3/12$, doubling the breadth of a rectangular beam decreases the deflection by one half, whereas doubling the depth of a beam decreases the deflection to one-eighth of the previous value.)

"Stiffness" of Material

The stiffer the material of a beam, i.e. the greater its resistance to bending, the less will be the deflection, other conditions such as span, load, etc., remaining constant. The measure of the "stiffness" of a material is its modulus of elasticity E and deflection is inversely proportional to the value of E.

DERIVATION OF DEFLECTION FORMULAE

A formula for calculating deflection must therefore contain: the load W, the cube of the span l^3, the moment of inertia I, and the modulus of elasticity E. For standard cases of loading, the deflection formula can be expressed in the form cWl^3/EI, where c is a numerical coefficient depending on the disposition of the load and also on the manner in which the beam is supported, that is, whether the ends of the beam are simply supported or fixed, etc. For Fig. 11.4 the values of c are respectively 1/48 and 5/384. W and l^3 are in the numerator of the formula because increase in their values means increase of deflection, whereas E and I are in the denominator because increase in their values means decrease of deflection.

Referring to Fig. 11.4 it should be obvious (neglecting the weights of the beams) that although the beams are equally loaded, the deflection of beam b will be less than that of beam a. In fact, the maximum deflection of beam a is

(a)

(b)

Figure 11.4

$$\frac{1}{48}\frac{Wl^3}{EI} = \frac{8}{384}\frac{Wl^3}{EI}$$

and the maximum deflection of beam b is

$$\frac{5}{384}\frac{Wl^3}{EI}$$

Table 11.1 gives the values of c for some common types of loading, etc. When the load system is complicated, e.g. several point loads of different magnitudes, or various combinations of point loads and uniformly distributed loads, the deflections must be calculated from first principles.

In certain simple cases it is possible to derive deflection formulae mathematically without using the calculus, and the

Table 11.1 Values of coefficient c for deflection formula $\delta = cWl^3/EI$

Condition of Loading	Value of c (δ_{max} at A)
$\frac{1}{2}l$ W $\frac{1}{2}l$ A	$\frac{1}{48}$ 0.02083
$\frac{1}{3}l$ $\frac{1}{2}W$ $\frac{1}{3}l$ $\frac{1}{2}W$ $\frac{1}{3}l$ A	$\frac{23}{1296}$ 0.01775
$\frac{1}{4}l$ $\frac{1}{2}W$ $\frac{1}{2}l$ $\frac{1}{2}W$ $\frac{1}{4}l$ A	$\frac{11}{768}$ 0.01432
A udl W l	$\frac{5}{384}$ 0.01302
$\frac{1}{2}l$ W $\frac{1}{2}l$ A Fixed Beam	$\frac{1}{192}$ 0.00521
l udl W Fixed Beam	$\frac{1}{384}$ 0.00260
l W A Cantilever	$\frac{1}{3}$ 0.33333
l udl W A Cantilever	$\frac{1}{8}$ 0.12500

Shear Force Diagram

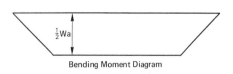

Bending Moment Diagram

Figure 11.5

Figure 11.6

Figure 11.7

following example is given for the more mathematically minded student. Neglecting the weight of the beam, Shear Force and Bending Moment Diagrams are given in Fig. 11.5 for the beam loaded as shown. The maximum bending moment is $Wa/2$ and this moment is constant along the length AB, the shear force being zero. Since the moment is constant, the portion AB of the beam bends into the arc of a circle with a radius of curvature R. In triangle OBC (Fig. 11.6),

$$R^2 = (R - \delta)^2 + (\tfrac{1}{2}l)^2$$
$$= R^2 - 2R\delta + \delta^2 + \tfrac{1}{4}l^2$$
$$R^2 - R^2 + 2R\delta - \delta^2 = \tfrac{1}{4}l^2$$
$$2R\delta - \delta^2 = \tfrac{1}{4}l^2 \tag{1}$$

δ^2 is the square of a small quantity and can be ignored. Therefore

$$2R\delta = \tfrac{1}{4}l^2 \tag{2}$$

For example, if $l = 3000$ mm and $\delta = 30$ mm, which is a big deflection for such a small span, then from equation (1)

$$2R \times 30 - 900 = \tfrac{1}{4} \times 9 \times 10^6$$
$$60R - 900 = 2.25 \times 10^6$$
$$60R = 2\,249\,100$$
$$R = 37\,485 \text{ mm}$$

If we ignore the term δ^2, $R = 37\,500$ mm, so the inaccuracy is very small, even with such a comparatively large deflection.

It was shown in Chapter 9, page 142 that

$$\frac{M}{I} = \frac{E}{R} \quad \text{or} \quad R = \frac{EI}{M}$$

Substituting this value of R in equation (2)

$$2 \times \frac{EI}{M} \times \delta = \tfrac{1}{4}l^2$$

$$\delta = \frac{Ml^2}{8EI}$$

M is the bending moment and equals $\tfrac{1}{2}Wa$. Therefore

$$\delta = \frac{Wal^2}{16EI} \tag{3}$$

a can be expressed in terms of the span l. If, for example, the overhanging portion a is a quarter ($\tfrac{1}{4}l$) of the interior span (Fig. 11.7) then, substituting in equation (3)

$$\delta = \frac{W \times \tfrac{1}{4}l \times l^2}{16EI} = \frac{1}{64}\frac{Wl^3}{EI}$$

Example 11.1

A 406×17.8 UB54 is simply supported at the ends of a span of 5 m and carries a uniformly distributed load of 60 kN/m. Calculate the maximum deflection ($E = 210\,000$ N/mm).

Solution The formula for the maximum deflection is

$$\delta = \frac{5}{384} \frac{Wl^3}{EI} \qquad\qquad \text{(Table 11.1)}$$

where

$$W = 60 \times 5 = 300 \text{ kN} = 0.3 \times 10^6 \text{ N}$$
$$l = 5000 \text{ mm}$$
$$E = 210\,000 \text{ N/mm}^2$$
$$I = 164.6 \times 10^6 \text{ mm}^4$$

Therefore

$$\delta = \frac{5}{384} \times \frac{300\,000 \times 125 \times 10^9}{210\,000 \times 164.6 \times 10^6}$$
$$= 14 \text{ mm}$$

Example 11.2

Calculate the safe inclusive uniformly distributed load for a 457×152 UB52 simply supported at its ends if *a*) the span is 6 m, *b*) the span is 12 m. The maximum permissible bending stress is 165 N/mm² and the maximum permissible deflection is 1/360 of the span. E is 210 000 N/mm².

Solution

a) $Z = 949\,000 \text{ mm}^3$ (from Table)
$$M_r = f \times Z = 165 \times 949\,000 = 156.6 \times 10^6 \text{ N mm}$$
$$M_{max} = \tfrac{1}{8}Wl = \tfrac{1}{8}W \times 6000 = 156.6 \times 10^6 \text{ N mm}$$

Therefore

$$W = \frac{156.6 \times 10^6 \times 8}{6000} = 208.8 \text{ kN}$$

Maximum deflection $\delta = \dfrac{5}{384} \dfrac{Wl^3}{EI}$

where $W = 208\,000 \text{ N}$
$$l = 6000 \text{ mm}$$
$$E = 210\,000 \text{ N/mm}^2$$
$$I = 190.4 \times 10^6 \text{ mm}^4 \quad \text{(from Table)}$$

Therefore

$$\delta = \frac{5}{384} \times \frac{208.8 \times 6^3 \times 10^{12}}{0.21 \times 190.4 \times 10^{12}} = 15 \text{ mm}$$

Maximum permissible deflection $= \dfrac{6000}{360} = 17 \text{ mm}$

Therefore

Safe load $= 208.8 \text{ kN}$

b) $M_r = 156.6 \times 10^6 \text{ N mm}$ (as above in a))

Therefore

$$W = \frac{156.6 \times 10^6 \times 8}{12\,000} = 104.4 \text{ kN}$$

Maximum actual deflection due to this load

$$= \frac{5}{384} \times \frac{104.4 \times 12^3 \times 10^{12}}{0.21 \times 190.4 \times 10^{12}} = 59 \text{ mm}$$

Maximum permissible deflection $= \dfrac{12\,000}{360} = 33 \text{ mm}$

This means that, although the beam is quite satisfactory from the strength point of view, the deflection is too great, therefore the load must be reduced.

Now $\quad \delta = \dfrac{5}{384} \dfrac{Wl^3}{EI}$

Therefore

$$33 = \frac{5}{384} \times \frac{W \times (12\,000)^3}{210\,000 \times 190.4 \times 10^6}$$

giving $\quad W = 58.6 \text{ kN}$

and this is the maximum permitted load for the beam. (Instead of using the deflection formula, W can also be obtained from $W = (33/59) \times 104.4 = 58.6 \text{ kN}$.)

Example 11.3

Calculate the safe inclusive uniformly distributed load for a 200 mm $\times$ 75 mm timber joist, simply supported at its ends, if a) the span is 4 m, b) the span is 8 m. The maximum permissible bending stress is 6 N/mm^2 and the maximum permissible deflection is 0.003 of the span. E is 9500 N/mm^2.

Solution

a) From $\quad M_r = \frac{1}{6}fbd^2 = M_{max} = \frac{1}{8}Wl$

$$\tfrac{1}{6} \times 6 \times 75 \times 200^2 = \tfrac{1}{8} \times W \times 4000$$

and so $W = 6.0 \text{ kN}$

$$\text{Maximum actual deflection} = \frac{5}{384} \frac{Wl^3}{EI}$$

where $W = 600 \text{ N}$

$\qquad l = 4000 \text{ mm}$

$\qquad E = 9500 \text{ N/mm}^2$

$\qquad I = \tfrac{1}{12} \times 75 \times 200^3 = 50 \times 10^6 \text{ mm}^4$

$$\delta = \frac{5}{384} \times \frac{6 \times 4^3 \times 10^{12}}{9.5 \times 50 \times 10^9} = 11 \text{ mm}$$

Maximum permissible deflection $= 0.003 \times 4000$

$$= 12 \text{ mm}$$

The safe udl for the 4 m span $= 6.0 \text{ kN}$.

b) In this case the span is doubled, therefore only half the previous load will be applied: $W = 3.0 \text{ kN}$.

Maximum actual deflection due to this load

$$= \frac{5}{384} \times \frac{3 \times 8^3 \times 10^{12}}{9500 \times 50 \times 10^6} = 42 \text{ mm}$$

Maximum permissible deflection $= 0.003 \times 8000 = 24 \text{ mm}$

The load must be reduced so that

$$24 = \frac{5}{384} \times \frac{Wl^3}{EI}$$

or the answer can be obtained by multiplying the load of 3 kN by 24/42 which gives 1.7 kN. Therefore, for this span, the safe load is 1.7 kN.

SPAN/DEPTH RATIOS

It appears from the above examples that, before a beam can be passed as suitable, deflection calculations must be made in addition to bending calculations. Fortunately, it is possible to derive simple rules which replace deflection calculations in the majority of cases. For example, if having designed a UB simply supported at its ends and carrying a udl ($f = 165 \text{ N/mm}^2$) it is found that the span of the beam does not exceed 17 times its depth, the beam will be suitable from the deflection point of view.

This rule is derived as follows.

The maximum deflection for a simply supported beam with a udl is

$$\delta = \frac{5}{384} \times \frac{Wl^3}{EI}$$

This formula can be re-arranged as follows,

$$\delta = \frac{5}{48} \times \frac{Wl}{8} \times \frac{l^2}{EI}$$

Now $\frac{1}{8}Wl$ is the maximum bending moment which is M. Thus

$$\delta = \frac{5}{48} \times \frac{M}{I} \times \frac{l^2}{E}$$

But $M/I = f/y$ (see page 142)

$$\delta = \frac{5}{48} \times \frac{f}{y} \times \frac{l^2}{E}$$

$f = 165$ N/mm^2, $y =$ half the depth of the beam $= \frac{1}{2}d$ (see page 143); and E is 210 000 N/mm^2. Therefore

$$\delta = \frac{5}{48} \times \frac{165 \times 2}{d} \times \frac{l^2}{210\,000} = \frac{1650l^2}{10.08 \times 10^6 d} = \frac{l^2}{6110d}$$

but δ must not exceed $l/360$, hence

$$\frac{l}{360} = \frac{l^2}{6110d}$$

$$360\,l = 6110d$$

$$\frac{l}{d} = \frac{6110}{360} = 17 \quad \text{or} \quad l = 17d$$

Similar rules can be derived in the same manner for rectangular timber sections. Table 11.2 gives values for the stresses and values of E as given in CP112:1972. These rules are only applicable to beams simply supported at each end and carrying a udl, but they err on the side of safety in the case of beams carrying a point load at mid-span.

Note that for each of the grades of timber mentioned in Table 11.2 two values of the elastic modulus are given. CP112 states that the mean value of the modulus of elasticity shall only be used for rafters, floor and ceiling joists, trusses and stud-walling. In all other cases, the minimum value of the modulus of elasticity shall be used.

Table 11.2 Span/depth ratios for timber beams with udl over full span and $\delta \not> 0.003 \times$ span

Species Group	Grade	f in N/mm^2	E in N/mm^2 Minimum	Mean	$\dfrac{l}{d}$
S1	SS	8.6	5100		8.5
				10600	17.7
	GS	6.0	4600		11.0
				9500	22.8
S2	SS	6.9	4800		10.0
				8900	18.6
	GS	4.8	4300		12.9
				8000	24.0

DEFLECTION OF REINFORCED CONCRETE BEAMS

The composite nature of reinforced concrete beams complicates the determination of their deflections and, although these may be calculated, the process is tedious and time consuming. For this reason, both CP114 and CP110 state that for "all normal cases" the beam will not deflect excessively provided the span/depth ratios are kept within the specified limits.

Table 11.3 shows some of those ratios. It should be noted that, whereas in CP114 the overall depth of the beam is used, CP110 stipulates the effective depth.

Table 11.3 Span/depth ratios for reinforced concrete beams

Type of beam	CP114	CP110
Cantilever	10	7
Simply supported	20	20
Continuous	25	26

SUMMARY

For beams which behave elastically (steel, timber) and for standard types of loading, the **actual deflection** is

$$\delta = c \times \frac{W}{E} \times \frac{l^3}{I}$$

where $c =$ a numerical coefficient taking into account the load system (udl, point load, etc.) and the manner in which the beam is supported (fixed or simple supports)

$W =$ the total load on the beam

$l =$ the span

$E =$ modulus of elasticity of the material

$I =$ moment of inertia (second moment of area) of the section.

Limitations are imposed on the maximum amount of deflection by the British Standard Specifications and Codes of Practice. In general,

for steel the limit is $(1/360) \times$ span

for timber the limit is $0.003 \times$ span.

Reinforced concrete beams (and slabs) are deemed to satisfy the limitation if their specified span/depth ratio is not exceeded.

EXERCISE 11

Note The value of E for steel is $210\,000\,\text{N/mm}^2$. The values of E for timber are given in Table 11.2.

1 A 457×191UB98 is simply supported at the ends of a span of 7.2 m. The beam carries an inclusive udl of 350 kN. Calculate the maximum deflection.

2 A 406×178UB60 is simply supported at the ends of a span of 6.0 m. The beam carries a point load of 140 kN at mid-span. Calculate the deflection due to this load, ignoring the weight of the beam.

3 A 356×171UB45 is simply supported at the ends of a span of 5.5 m. It carries an inclusive udl of 12 kN/m and a central point load of 75 kN. Calculate the maximum deflection.

4 Calculate the safe, inclusive udl for a 356×171UB67 simply supported at the ends of a span of 9 m. The permissible bending stress is 165 N/mm² and the deflection must not exceed $(1/360) \times$ span.

5 Calculate the maximum deflection of a 305×165UB40 cantilevered 3 m beyond its fixed support and carrying an inclusive udl of 30 kN.

6 A 203×133UB30 is fixed at one end and cantilevers for a distance of 1.2 m. The beam supports a point load of 15 kN at its free end. Calculate the maximum deflection ignoring the weight of the beam.

7 A 254×146UB43 is fixed at one end and cantilevers for a distance of 1.5 m. The beam carries a udl of 12 kN/m and a point load of 10 kN at the free end. Calculate the maximum deflection.

8 A 75 mm wide and 150 mm deep beam in S2/SS grade timber carries an inclusive udl over a simply supported span of 1.8 m. Calculate the maximum deflection for udl of 10 kN.

9 A timber (S2/GS) beam 75 mm wide and 240 mm deep carries a point load of 5.5 kN at the centre of its simply supported span of 3 m. Calculate the maximum deflection due to this load.

10 A timber (S1/GS) beam, 100 mm × 300 mm, spans 4 m and carries a central point load of 5 kN in addition to an inclusive udl of 10 kN. Calculate the maximum deflection.

11 A 75 mm × 225 mm S1/SS timber beam spans 5 m on simple supports. Calculate the value of the safe udl for the following conditions:
a) permissible bending stress of 8.6 N/mm^2 and deflection is not important;
b) the deflection must not exceed 0.003 × span.

12 A beam in S1/GS timber carries a udl of 7.8 kN on a simply supported span of 5 m. Assuming the beam to be 75 mm wide calculate its depth when
a) the permissible bending stress is 6 N/mm^2 and deflection is not important;
b) maximum deflection is limited to 0.003 × span.

13 Calculate the maximum deflection of a cantilever beam in S2/SS timber 75 mm wide and 240 mm deep. The beam carries an inclusive udl of 5 kN over a span of 2 m.

14 A timber (S2/GS) cantilever beam is 1.2 m long, 75 mm wide and 150 mm deep. It carries a udl of 1.0 kN and a point load of 0.8 kN at its free end. Calculate the maximum deflection.

15 Floor joists in S2/GS timber spaced at 360 mm c/c are to carry an inclusive load of 2.0 kN/m^2 over a simply supported span of 4.0 m.
  Determine a suitable size for the joists if the bending stress is limited to 4.8 N/mm^2 and the deflection must not exceed 0.003 × span.

16 Calculate the maximum span/depth ratio for a steel cantilever beam supporting a udl so that deflection does not exceed (1/360) × span. The permissible bending stress is 155 N/mm^2.

17 Calculate the minimum depth for a steel beam simply supported at the ends of a 7.5 m span carrying a point load at mid-span. The weight of the beam may be ignored. The deflection must be limited to the usual (1/360) × span and the permissible bending stress is 165 N/mm^2.

18 Determine the deflection of a 150 mm × 400 mm S2/GS timber beam which is simply supported at the ends of its 5 m length. The beam is subjected to a maximum bending moment of 24×10^6 N mm due to an inclusive udl.

19 A solid rectangular beam is simply supported at the ends of a 4.8 m span. The beam is subjected to a maximum bending moment of 90×10^6 N mm due to an inclusive udl.

Determine a suitable section for the beam given:

i) modulus of elasticity of the material = 4000 N/mm^2;

ii) maximum permissible bending stress = 7.5 N/mm^2;

iii) the deflection must be limited to 15 mm.

20 Calculate the minimum depth for a simply supported S2/SS timber beam, which is to carry a udl over a span of 4.0 m, if the deflection must not exceed $0.002 \times$ span and the permissible stress in bending is 6.9 N/mm^2.

Assuming the beam to be 240 mm wide determine the value of the udl.

21 A timber beam 75 mm × 150 mm simply supported at the ends of a 2.0 m span deflects 5 mm under a 10 kN udl. Without calculating the modulus of elasticity, determine the maximum deflection of a beam of similar timber 150 mm × 300 mm due to a udl of 40 kN on a span of 4.0 m.

12 Axially Loaded Columns

Figure 12.1

Figure 12.2

In this chapter the factors affecting the column's load carrying capacity are investigated. The connection between the slenderness of the column and its tendency to buckle is discussed. The influence of the "fixity" of the ends of the column, and the shape of its section on that slenderness, is considered in relation to timber, steel and reinforced concrete columns.

When the line of action of the resultant load is coincident with the centre of gravity axis of the column (Fig. 12.1), the column is said to be **axially loaded** and the stress produced in the material is said to be a *direct compressive stress*. This stress is uniform over the cross-section of the column. The term *concentric loading* is sometimes used instead of *axial loading*.

When the load is not axial, it is said to be eccentric (i.e. off-centre) and bending stress is induced in the column as well as a direct compressive stress (Fig. 12.2). Eccentric loading is dealt with in Chapter 14.

Other words used to describe members which are subjected to compressive stress are *pillar, post, stanchion, strut.* There are no definite rules as to when any one of these words should be used, but the following convention is fairly general.

Column and pillar can usually be applied to any material, e.g. timber, stone, concrete, reinforced concrete, steel. Post is usually confined to timber. Stanchion is often used for rolled steel I-sections and channel sections. Strut has a more general significance than stanchion or post but normally it is not applied to a main supporting member of a building. The word is often used for compression members of roof trusses whether the material is timber or steel.

Figure 12.3

DESIGN FACTORS

The maximum axial load a column can be allowed to support depends on

1 The material of which the column is made.

2 The slenderness of the column, which involves not only the height or length of the column, but also the size and shape of its cross-section and the manner in which the two ends of the column are supported or fixed.

The majority of columns are designed by reference to tables of permissible stresses contained in British Standard Specifications and Codes of Practice. These tables of permissible stresses (which are reproduced on pages 196 and 200) have been constructed from complex formulae which have been derived as the result of a great deal of research, mathematical and experimental, into the behaviour of columns under load. It is not possible in this book to deal with the mathematical theories of column design but an attempt will be made to give an explanation of the general principles.

A very short column will fail due to crushing of the material, but long columns are likely to fail by "buckling", the failing load being much less than that which would cause failure in a short column of identical cross-sectional dimensions.

Consider Fig. 12.3, which represents a strip of pliable wood, say 6 mm × 54 mm in cross-section and 600 mm long. A small vertical force applied as shown will cause buckling. It should be obvious (but can be confirmed by experiment if necessary) that a larger force would be required to cause failure if the member were only 300 mm high. In other words, the 600 mm high member is more slender than the 300 mm high member of equal cross-sectional dimensions.

Now consider Fig. 12.4. All three members are of equal cross-sectional area (324 mm²) and of equal height, yet member b will require more load to cause buckling than member a and member c is the strongest column.

Slenderness Ratio

By reference to Fig. 12.4 it can be seen that the smaller cross-sectional dimension of the column is very important from the point of view of buckling and it appears that the **slenderness ratio** of a column can be defined as

$$\frac{\text{Effective length of column (in millimetres)}}{\text{Least width of column (in millimetres)}} = \frac{l}{b}$$

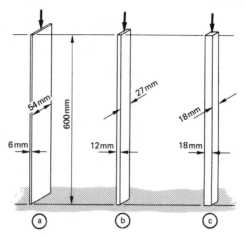

Figure 12.4

e.g. slenderness ratio of

Column $a = 600/6 = 100$

Column $b = 600/12 = 50$

Column $c = 600/18 = 33$ approx.

Since most timber posts and struts are rectangular in cross-section, it is reasonable to express the slenderness ratio in terms of the length and least width (i.e. least lateral dimension). This method is, in fact, adopted by CP112 (Table 12.1) but an alternative slenderness ratio is also given, i.e.

$$\frac{\text{Effective length}}{\text{Least radius of gyration}} = \frac{l}{r}$$

and it is this slenderness ratio which must be used when the post is not of solid rectangular cross-section.

A Swiss mathematician named Leonhard Euler (1707–1783) showed that a long thin homogeneous column, axially loaded, suffers no deflection as the load is gradually applied until a critical load (the collapsing or buckling load P) is reached. At this load, instability occurs and the column buckles into a curve.

The curve of Fig. 12.5 is not the arc of a circle, and Euler found (with the aid of the calculus) that the buckling load P gets less as the slenderness of the column increases.

Euler's formula is not used for design, since (except for very long columns) it gives a value of the collapsing load which is much higher than the actual collapsing load of practical columns, but it still forms part of modern column formulae.

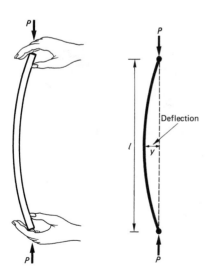

Figure 12.5

Table 12.1 Maximum Permissible Compressive Stresses in N/mm² for Struts in Group S2 Timber (adapted from Tables 11A and 15 of CP112)

$\dfrac{l}{r}$	$\dfrac{l}{b}$	Grades		$\dfrac{l}{r}$	$\dfrac{l}{b}$	Grades	
		SS	GS			SS	GS
<5	1.4	7.0	4.8	90	26.0	4.27	2.93
5	1.4	6.93	4.75	100	28.8	3.71	2.54
10	2.9	6.86	4.70	120	34.6	2.80	1.92
20	5.8	6.72	4.61	140	40.4	2.17	1.49
30	8.7	6.58	4.51	160	46.2	1.68	1.15
40	11.5	6.37	4.37	180	52.0	1.40	0.96
50	14.4	6.09	4.18	200	57.7	1.12	0.77
60	17.3	5.81	3.98	220	63.5	0.91	0.62
70	20.2	5.39	3.70	240	69.2	0.77	0.53
80	23.0	4.90	3.36	250	72.2	0.70	0.48

Example 12.1

Calculate the permissible axial load for the following timber posts of S2/GS grade timber, all the posts having an effective length of 3.45 m:

a) 150 mm × 150 mm   *b*) 225 mm × 100 mm
c) 300 mm × 75 mm

(Note that these posts have equal cross-sectional areas.)

Solution
a) Slenderness ratio = 3450/150 = 23
From Table 12.1

Permissible stress = 3.36 N/mm²

Area of cross-section = 22 500 mm²

Permissible axial load = 3.36 × 22 500 = 75.6 kN

b) Slenderness ratio = 3450/100 = 34.5
By interpolation,

Permissible stress for *l*/*b* of 34.5 = 1.93 N/mm²

Permissible axial load = 1.93 × 22 500 = 43.4 kN

c) Slenderness ratio = 3450/75 = 46
Again, by interpolation,

Permissible stress = 1.16 N/mm²

Permissible axial load = 1.16 × 22 500 = 26.1 kN

It is interesting to note that the 150 mm × 150 mm post can carry almost three times the load permitted for the 300 mm × 75 mm post.

Example 12.2

A timber post of S2/GS grade timber of 3.45 m effective length is required to support an axial load of 42 kN. Determine suitable dimensions for the cross-section of the post.

Solution Dimensions must be assumed because it is not possible to determine the permissible stress until the slenderness ratio is known. For example, let the first trial be a post 150 mm square (area of cross-section = 22 500 mm²).

Actual stress due to the load = 42 000/22 500

$$= 1.87 \text{ N/mm}^2$$

Slenderness ratio = l/b = 3450/150 = 23

The permissible stress for a slenderness ratio of 23 is 3.36 N/mm² therefore the assumed size of 150 mm square is too large. As a second trial, assume a post 125 mm square.

Actual stress = 42 000/15 625 = 2.69 N/mm²

Slenderness ratio = l/b = 3450/125 = 27.6

From Table 12.1

Permissible stress for l/b of 26 = 2.93 N/mm²

for l/b of 28.8 = 2.54 N/mm²

Difference in stress for 2.8 = 0.39

Difference in stress for 1.6 = 0.22

Permissible stress for l/b of 27.6 = 2.93 − 0.22

$$= 2.71 \text{ N/mm}^2$$

Since the permissible stress is slightly more than the actual stress due to the load, a 125 mm square post is suitable.

Radius of Gyration

It has already been stated that the shape of the cross-section has an important influence on the load carrying capacity of a column. A square timber post can support more load than a post of rectangular cross-section of equal area and equal height. It has been shown also that in the case of rectangular cross-section columns it is reasonable to base permissible stresses on a slenderness ratio obtained by dividing the length of the column by its least width. It is not possible, however, to use the dimension of least width when designing columns which have other than solid rectangular sections. Steel columns, for example, are made in various shapes, some of which are shown in Fig. 12.6.

Figure 12.6

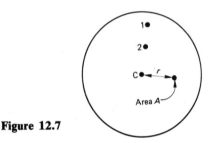

Figure 12.7

For such columns it is necessary to use some method of calculating slenderness ratios which can be applied to any shape of cross-section. A property which takes into account not only the size of the section (i.e. area) but also its shape (i.e. the arrangement of the material in the cross-section) is the **radius of gyration.** It is obtained by dividing the moment of inertia I of the section by its area A and then extracting the square root. The symbol r is commonly used to denote radius of gyration but k and g are sometimes used.

$$\text{Radius of gyration } r = \sqrt{\left(\frac{I}{A}\right)}$$

The use of the word "gyration" when applied to stationary columns in buildings may appear strange until it is realized that the word is also used in dynamics, the branch of mechanics dealing with bodies in motion. For example, consider Fig. 12.7 which represents a disc (such as a flywheel) rotating about its centre C. In dynamics, it is usually the mass (weight) of the wheel which enters into calculations but in this instance the area A of the disc will be considered. Different particles of the disc travel at different velocities. For example, in one revolution of the wheel, particle 1 travels a greater distance than particle 2. In estimating the total energy of the disc, the term $\sum ay^2$ enters into the calculations. $\sum ay^2$ is the sum of the second moments of all the particles of area about the centre C of the disc, i.e. the moment of inertia I.

Imagine that all the area A of the disc is concentrated into *one* imaginary particle at a distance r from the centre of the disc. The moment of inertia about C of this particle is Ar^2 and, if the total energy of the disc is to remain unaltered, Ar^2 must equal the total moment of inertia I of the disc. Therefore

$$Ar^2 = I$$

and

$$r = \sqrt{\left(\frac{I}{A}\right)}$$

The radius of gyration is, in this connection, the distance from the centre C of the disc to the point at which the whole area of the wheel can be assumed to be concentrated so that the total energy remains unaltered.

In structural work, it is convenient to use the property $\sqrt{(I/A)}$ in conjunction with the length of the column for estimating slenderness ratios.

Least Radius of Gyration

The structural engineer is not concerned with moment of inertia about a point as in the disc discussed above, but with moment of inertia with reference to a given axis. If a column of I section buckles under its load, the bending will be about the weaker axis (axis Y–Y), as indicated in Fig. 12.8. Therefore, the radius of gyration must be calculated from I_{yy} which is the least moment of inertia.

Figure 12.8

$$\text{Least radius of gyration} = \sqrt{\left(\frac{\text{Least moment of inertia}}{\text{Area of cross-section}}\right)}$$

i.e. $\quad r = \sqrt{\left(\frac{I_{yy}}{A}\right)}$

$$\text{Slenderness ratio} = \frac{\text{Effective length of column (mm)}}{\text{Least radius of gyration (mm)}}$$

$$= \frac{l}{r}$$

Example 12.3

A timber post is $150\,\text{mm} \times 100\,\text{mm}$ in cross-section and has an effective length of 2.6 m. Calculate its least radius of gyration and the slenderness ratio.

Figure 12.9

Solution
From Fig. 12.9,

least $I = I_{yy} = db^3/12$ and the area $A = db$

Therefore

$$\text{least } r = \sqrt{\left(\frac{I}{A}\right)} = \sqrt{\left(\frac{db^3}{12} \times \frac{1}{ab}\right)} = \sqrt{\left(\frac{b^2}{12}\right)} = \frac{b}{\sqrt{12}} = 0.289b$$

$r_{yy} = 0.289b$

(Note that $r_{xx} = 0.289d$)

Least $r = 0.289b = 0.289 \times 100 = 28.9\,\text{mm}$

Effective length $= 2600\,\text{mm}$

Slenderness ratio $= l/r = 2600/28.9 = 90$

By reference to Table 12.1 it will be seen that l/r of 90 corresponds to l/b of 26.0; and in this case $l/b = 2600/100 = 26.0$. Thus, when designing timber posts of solid rectangular cross-section by reference to Table 12.1, it is immaterial whether the slenderness ratio is taken as effective height divided by least radius of gyration (l/r) or as effective height divided by least width (l/b).

No such dilemmas arise, however, in the design of steel columns. Because of their varied shapes the allowable com-

Table 12.2 Allowable Compressive Stress p_c on Gross Section for Axial Loading (from Table 17a BS449:Part 2:1969)

$\dfrac{l}{r}$	p_c (N/mm^2) for grade 43 steel									
	0	1	2	3	4	5	6	7	8	9
0	155	155	154	154	153	153	153	152	152	151
10	151	151	150	150	149	149	148	148	148	147
20	147	146	146	146	145	145	144	144	144	143
30	143	142	142	142	141	141	141	140	140	139
40	139	138	138	137	137	136	136	136	135	134
50	133	133	132	131	130	130	129	128	127	126
60	126	125	124	123	122	121	120	119	118	117
70	115	114	113	112	111	110	108	107	106	105
80	104	102	101	100	99	97	96	95	94	92
90	91	90	89	87	86	85	84	83	81	80
100	79	78	77	76	75	74	73	72	71	70
110	69	68	67	66	65	64	63	62	61	61
120	60	59	58	57	56	56	55	54	53	53
130	52	51	51	50	49	49	48	48	47	46
140	46	45	45	44	43	43	42	42	41	41
150	40	40	39	39	38	38	38	37	37	36
160	36	35	35	35	34	34	33	33	33	32
170	32	32	31	31	31	30	30	30	29	29
180	29	28	28	28	28	27	27	27	26	26
190	26	26	25	25	25	25	24	24	24	24
200	24	23	23	23	23	22	22	22	22	22
210	21	21	21	21	21	20	20	20	20	20
220	20	19	19	19	19	19	19	18	18	18
230	18	18	18	18	17	17	17	17	17	17
240	17	16	16	16	16	16	16	16	16	15
250	15									
300	11									
350	8									

Intermediate values may be obtained by linear interpolation.

Note: For material over 40 mm thick, other than rolled I-beams or channels, and for Universal columns of thicknesses exceeding 40 mm, the limiting stress is 140 N/mm².

pressive stress p_c, shown in Table 12.2, is based on the l/r slenderness ratio.

Effective Length of Columns

It should be noted that in discussing the slenderness ratio the length of the column was qualified by the term "effective". This is in accordance with the provisions of BS449 and the relevant Codes of Practice which state that, for the purpose of calculating the slenderness ratio of columns, an effective length should be assumed. This **effective length** can be defined as that length of the column which is subject to buckling.

BS449 and the relevant Codes of Practice give guidance on the relationship of the length of the column between lateral supports, L, to its effective length l. This is summarised in Table 12.3.

The reason why the effective length of a column may be less than or greater than its actual length in a building or structure is as follows. The safe compressive stress for a column depends not only on the actual length and cross-sectional dimensions of the column but also on the manner in which the ends of the column are restrained or fixed. Tables

Table 12.3 Effective Length of Columns

Type of "fixity"	The effective length l of column BS449 CP112	CP114 CP110
1 Effectively held in position and restrained in direction at both ends.	0.7L	0.75L
2 Effectively held in position at both ends and restrained in direction at one end.	0.85L	
3 Effectively held in position at both ends, but not restrained in direction.	L	0.75L − L
4 Effectively held in position and restrained in direction at one end, and at the other partially restrained in direction but not held in position.	1.5L	
5 Effectively held in position and restrained in direction at one end, but not held in position or restrained in direction at the other end.	2.0L	L − 2.0L

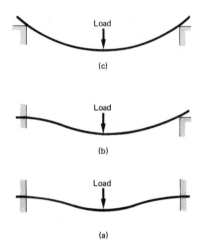

Figure 12.10 (*a*) Both ends of beam freely supported (bending exaggerated; (*b*) One end of beam fixed, one end freely supported; (*c*) Both ends fixed

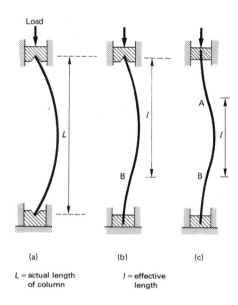

L = actual length of column

l = effective length

Figure 12.11 All column ends are held in position: (*a*) Top and bottom of column not restrained in direction; (*b*) Top not restrained in direction; (*c*) Top and bottom restrained in direction

12.1 and 12.2 have been derived for one condition of end fixing (both ends pinned or hinged). To make allowance for other conditions of end fixing, instead of constructing further tables of permissible stresses, adjustment is made by using a different length of column when calculating the slenderness ratio.

Before considering the end fixing of columns, it may be instructive to study the behaviour of beams as indicated in Fig. 12.10 (*a*), (*b*) and (*c*).

In Fig. 12.10 (*a*) both ends of the beam are free to bend upwards when the load is applied; in other words, the ends of the beam are not restrained in direction, although they are held in position.

In Fig. 12.10 (*b*) one end of the beam is firmly fixed so that the end is restrained in direction, and in Fig. 12.10 (*c*) both ends of the beam are restrained in direction.

The load carrying capacities of these three beams will be different (other conditions being equal) because of the manner in which the ends of the beams are held or supported.

In a comparable manner, columns will buckle into differently shaped curves according to the way in which the ends of the columns are held.

Side sway

Effective length *l*

Figure 12.12

Figure 12.13

The columns of Fig. 12.11 are all held in position, i.e. the ends of the columns are not free to move sideways or backwards or forwards. A greater load is required to cause column *b* to fail than column *a*, and column *c* is the strongest. The length of curve marked AB in columns *b* and *c* is similar to the whole length of curve of column *a*, and this length is called the effective length of the column.

Column *a* is similar to case (3) of Table 12.3 on page 201, column *b* is similar to case (2), and column *c* is similar to case (1).

Case (5) in Table 12.3 can be explained by reference to Fig. 12.12. The tops of the columns are not restrained either in position or direction and will tend to buckle as shown. The effective length is therefore taken as twice the actual length of the column.

Fig. 12.13 represents columns which are held in position and restrained in direction at the bottom and only restrained in direction at the top. This case is a little better than case (4) of Table 12.3.

Practical Interpretation

Appendix D to BS449:1969 gives typical examples of stanchions in multiple and single storey buildings and the effective lengths which may be used in their design. Attention is also drawn to the fact that, although Tables 12.1 and 12.2 give values of permissible stresses for slenderness ratios up to 250 and 350 respectively, it is only in special cases that the slenderness ratio is allowed to exceed 180. Clauses 3.14.2 of CP112:Part 2:1971 and 33 of BS449:Part 2:1969 should be consulted for details.

Example 12.4

Calculate the safe axial load for a 203×203UC60 stanchion which is 3.6 m high (between lateral supports). Both ends of the stanchion are held effectively in position but only one is also restrained in direction.

Solution The end fixity of the stanchion corresponds to case (2) of Table 12.3, so

Effective length of the column, $l = 0.85L = 3060 \text{ mm}^2$

From Table 12.5.

$r_{yy} = 51.9 \text{ mm} \qquad A = 7580 \text{ mm}^2$

Therefore

$l/r_{yy} = 3060/51.9 = 59$

From Table 12.2

p_c for l/r_{yy} of $59 = 126 \text{ N/mm}^2$

Hence,

Safe axial load $= p_c \times A$
$$= 126 \times 7580 = 955 \text{ kN}$$

Example 12.5

A stanchion of I section is required to support an axial load of 1000 kN. The effective length of the stanchion is 4 m. Choose a suitable section.

Solution From Table 12.4 it can be seen that a suitable section is a $254 \times 254 \text{UC}73$ stanchion which is capable of supporting an axial load of 1149 kN. When tables are not available it is necessary to make a guess at the size of the stanchion. For example, assuming a $203 \times 203 \text{UC}60$ stanchion and, working from first principles, the safe axial load is 812 kN and this stanchion is therefore not adequate. Another guess would then be made and so on until a suitable stanchion is found.

Try the $254 \times 254 \text{UC}73$. Then, from Table 12.5

$r_{yy} = 64.6 \text{ mm} \qquad A = 9290 \text{ mm}^2$

Therefore

$l/r_{yy} = 4000/64.6 = 62$

From Table 12.2 $\quad p_c = 124 \text{ N/mm}^2$

Safe axial load $= p_c \times A$
$$= 124 \times 9290 = 1152 \text{ kN}$$

Example 12.6

A hollow square column is made of 6 mm thick grade 43 steel and its external dimensions are 120 mm × 120 mm. Calculate the safe axial load if the effective length of the column is 3.5 m.

Solution
Radius of gyration $r = \sqrt{(I/A)}$

where $\quad I = \frac{1}{12}(BD^3 - bd^3) \quad$ and $\quad A = BD - bd$
$$= \tfrac{1}{12}(120^4 - 108^4) \qquad\quad = 120^2 - 108^2$$
$$= 5.94 \times 10^6 \text{ mm}^4 \qquad = 2736 \text{ mm}^2$$
$$r = 5\,940\,000/2736 = 46.6 \text{ mm}$$

Table 12.4 Universal columns as stanchions: safe loads for Grade 43 steel (from BCSA Structural Steelwork Handbook)

Serial size	Mass per metre	Safe concentric loads in kilonewtons for effective lengths in metres												
mm	kg	1.5	2.0	2.5	3.0	3.5	4.0	5.0	6.0	8.0	10.0	12.0	14.0	16.0
356 × 406	634	**11313**	**11313**	**11313**	**11313**	**11313**	**11313**	11002	10517	9090	7287	5643	4382	3462
	551	**9826**	**9826**	**9826**	**9826**	**9826**	**9826**	9527	9087	7801	6206	4782	3705	2923
	467	**8337**	**8337**	**8337**	**8337**	**8337**	**8337**	8057	7668	6534	5158	3955	3056	2408
	393	**7012**	**7012**	**7012**	**7012**	**7012**	7001	6755	6414	5427	4253	3247	2503	1970
	340	**6057**	**6057**	**6057**	**6057**	**6057**	6040	5821	5518	4644	3620	2755	2121	1668
	287	5455	5382	5309	5237	5174	5102	4911	4647	3889	3015	2288	1759	1381
	235	4467	4406	4346	4286	4235	4174	4013	3790	3153	2432	1840	1412	1108
Column Core	477	**8501**	**8501**	**8501**	**8501**	**8501**	8494	8201	7795	6615	5200	3978	3070	2417
356 × 368	202	3832	3778	3723	3671	3621	3560	3398	3173	2555	1917	1430	1090	852
	177	3353	3305	3257	3211	3168	3114	2970	2770	2224	1665	1241	945	739
	153	2899	2857	2815	2776	2737	2690	2563	2388	1909	1426	1061	807	631
	129	2448	2413	2377	2344	2311	2270	2161	2010	1601	1192	885	673	526
305 × 305	283	**5046**	**5046**	**5046**	**5046**	4964	4840	4505	4048	2981	2116	1539	1158	
	240	4507	4431	4356	4288	4200	4091	3796	3397	2480	1753	1272	957	
	198	3718	3654	3593	3534	3459	3366	3113	2772	2006	1412	1023	768	
	158	2962	2911	2862	2813	2751	2673	2462	2180	1562	1095	792	594	
	137	2570	2524	2481	2438	2383	2314	2126	1877	1338	935	676	507	
	118	2204	2164	2127	2090	2041	1981	1816	1599	1134	791	571		
	97	1813	1780	1749	1718	1677	1626	1487	1304	920	640	461		
254 × 254	167	3100	3036	2980	2905	2809	2686	2363	1977	1307	886	632		
	132	2445	2394	2348	2286	2206	2104	1838	1526	1000	676			
	107	1991	1949	1910	1858	1790	1704	1480	1222	796	537			
	89	1659	1624	1592	1547	1490	1417	1227	1010	656	442			
	73	1351	1323	1295	1259	1210	1149	991	813	525	354			
203 × 203	86	1580	1541	1488	1416	1321	1205	950	728	443				
	71	1306	1273	1229	1168	1087	990	777	594	360				
	60	1086	1058	1020	966	896	812	632	481	291				
	52	951	927	892	845	783	708	550	418	252				
	46	842	820	789	745	689	622	481	365	220				
152 × 152	37	661	626	574	505	429	357	249	180					
	30	533	504	460	403	341	283	197	142					
	23	413	388	351	303	253	208	144	103					

The above safe loads are tabulated for ratios of slenderness not exceeding 180. Values in ordinary type are calculated in accordance with Table 17a of BS 449. Values in bold type are based on the limiting stress of 140 N/mm^2 for material over 40 mm thick.

Table 12.5 Universal columns: dimensions and properties (from BCSA Structural Steelwork Handbook)

Serial size	Mass per metre	Depth of section	Width of section	Thickness		Root radius	Depth between fillets	Area of section
		D	**B**	Web **t**	Flange **T**	**r**	**d**	
mm	kg	mm	mm	mm	mm	mm	mm	cm²
356 × 406	634	474.7	424.1	47.6	77.0	15.2	290.1	808.1
	551	455.7	418.5	42.0	67.5	15.2	290.1	701.8
	467	436.6	412.4	35.9	58.0	15.2	290.1	595.5
	393	419.1	407.0	30.6	49.2	15.2	290.1	500.9
	340	406.4	403.0	26.5	42.9	15.2	290.1	432.7
	287	393.7	399.0	22.6	36.5	15.2	290.1	366.0
	235	381.0	395.0	18.5	30.2	15.2	290.1	299.8
Column Core	477	427.0	424.4	48.0	53.2	15.2	290.1	607.2
356 × 368	202	374.7	374.4	16.8	27.0	15.2	290.1	257.9
	177	368.3	372.1	14.5	23.8	15.2	290.1	225.7
	153	362.0	370.2	12.6	20.7	15.2	290.1	195.2
	129	355.6	368.3	10.7	17.5	15.2	290.1	164.9
305 × 305	283	365.3	321.8	26.9	44.1	15.2	246.5	360.4
	240	352.6	317.9	23.0	37.7	15.2	246.5	305.6
	198	339.9	314.1	19.2	31.4	15.2	246.5	252.3
	158	327.2	310.6	15.7	25.0	15.2	246.5	201.2
	137	320.5	308.7	13.8	21.7	15.2	246.5	174.6
	118	314.5	306.8	11.9	18.7	15.2	246.5	149.8
	97	307.8	304.8	9.9	15.4	15 2	246.5	123.3
254 × 254	167	289.1	264.5	19.2	31.7	12.7	200.2	212.4
	132	276.4	261.0	15.6	25.3	12.7	200.2	168.9
	107	266.7	258.3	13.0	20.5	12.7	200.2	136.6
	89	260.4	255.9	10.5	17.3	12.7	200.2	114.0
	73	254.0	254.0	8.6	14.2	12.7	200.2	92.9
203 × 203	86	222.3	208.8	13.0	20.5	10.2	160.8	110.1
	71	215.9	206.2	10.3	17.3	10.2	160.8	91.1
	60	209.6	205.2	9.3	14.2	10.2	160.8	75.8
	52	206.2	203.9	8.0	12.5	10.2	160.8	66.4
	46	203.2	203.2	7.3	11.0	10.2	160.8	58.8
152 × 152	37	161.8	154.4	8.1	11.5	7.6	123.4	47.4
	30	157.5	152.9	6.6	9.4	7.6	123.4	38.2
	23	152.4	152.4	6.1	6.8	7.6	123.4	29.8

Table 12.5 (Continued)

Serial size	Moment of inertia			Radius of gyration		Elastic modulus		Ratio
	Axis **x–x**		Axis	Axis	Axis	Axis	Axis	$\frac{D}{T}$
	Gross	Net	y–y	**x–x**	**y–y**	**x–x**	**y–y**	
mm	cm⁴	cm⁴	cm⁴	cm	cm	cm³	cm³	
356 × 406	275140	243065	98211	18.5	11.0	11592	4632	6.2
	227023	200308	82665	18.0	10.9	9964	3951	6.7
	183118	161334	67905	17.5	10.7	8388	3293	7.5
	146765	129152	55410	17.1	10.5	7004	2723	8.5
	122474	107674	46816	16.8	10.4	6027	2324	9.5
	99994	87837	38714	16.5	10.3	5080	1940	10.8
	79110	69430	31008	16.2	10.2	4153	1570	12.6
Column Core	172391	152946	68056	16.8	10.6	8075	3207	8.0
356 × 368	66307	57805	23632	16.0	9.57	3540	1262	13.9
	57153	49791	20470	15.9	9.52	3104	1100	15.5
	48525	42263	17469	15.8	9.46	2681	943.8	17.5
	40246	35047	14555	15.6	9.39	2264	790.4	20.3
305 × 305	78777	66878	24545	14.8	8.25	4314	1525	8.3
	64177	54405	20239	14.5	8.14	3641	1273	9.3
	50832	43039	16230	14.2	8.02	2991	1034	10.8
	38740	32783	12524	13.9	7.89	2368	806.3	13.1
	32838	27782	10672	13.7	7.82	2049	691.4	14.7
	27601	23344	9006	13.6	7.75	1755	587.0	16.8
	22202	18776	7268	13.4	7.68	1442	476.9	20.0
254 × 254	29914	27171	9796	11.9	6.79	2070	740.6	9.1
	22575	20492	7519	11.6	6.68	1634	576.2	10.9
	17510	15889	5901	11.3	6.57	1313	456.9	13.0
	14307	12973	4849	11.2	6.52	1099	378.9	15.0
	11360	10299	3873	11.1	6.46	894.5	305.0	17.9
203 × 203	9462	8373	3119	9.27	5.32	851.5	298.7	10.8
	7647	6756	2536	9.16	5.28	708.4	246.0	12.4
	6088	5383	2041	8.96	5.19	581.1	199.0	14.8
	5263	4651	1770	8.90	5.16	510.4	173.6	16.5
	4564	4035	1539	8.81	5.11	449.2	151.5	18.5
152 × 152	2218	1931	709	6.84	3.87	274.2	91.78	14.0
	1742	1516	558	6.75	3.82	221.2	73.06	16.8
	1263	1104	403	6.51	3.68	165.7	52.95	22.3

In calculating the net moment of inertia, each flange of 300 mm or greater width is reduced by two holes, and each flange less than 300 mm wide by one hole.

Therefore

$$l/r = 3500/46.6 = 75$$
$$p_c = 110 \text{ N/mm}^2$$

$$
\begin{aligned}
\text{Safe axial load} &= p_c \times A \\
&= 110 \times 2736 \\
&= 300 \text{ kN}
\end{aligned}
$$

REINFORCED CONCRETE COLUMNS

In the design of reinforced concrete columns it is necessary to distinguish between short and slender columns.

A short column, according to Clause 322.a.(i) of CP114, is one where the ratio of the effective length to least lateral dimension does not exceed 15. CP110 has a somewhat different ratio and will not be considered here.

The permissible direct compression stresses for nominal concrete mixes for short columns are as follows (Clause 303.a.(i) CP114):

$1:1:2$ mix	7.6 N/mm^2
$1:1\frac{1}{2}:3$	6.5 N/mm^2
$1:2:4$	5.3 N/mm^2

For longitudinal reinforcement, hot rolled steel bars complying with BS4449 may be used. The permissible stress for these bars up to 40 mm dia is 125 N/mm^2 (Clause 304.c. CP114).

For slender columns a reduction coefficient has to be used (see Clause 322.a.(iv) and Table 18 in CP114:Part 2:1969).

The permissible axial load P_0 for a short reinforced concrete column is calculated from the following equation:

$$P_0 = p_{cc}A_c + p_{sc}A_{sc}$$

where p_{cc} = the permissible stress for the concrete in direct compression;

A_c = the cross-sectional area of concrete excluding any finishing material and reinforcing steel;

p_{sc} = the permissible compression stress for column bars;

A_{sc} = the cross-sectional area of the longitudinal steel.

Example 12.7

A 250 mm square reinforced concrete column with an effective length of 3 m contains four 25 mm dia. longitudinal bars. Calculate the safe axial load for the column if $p_{cc} = 5.3 \text{ N/mm}^2$ and $p_{sc} = 125 \text{ N/mm}^2$.

Solution

$$\text{Slenderness ratio} = \frac{\text{Effective length}}{\text{Least width}}$$

$$= 300/250 = 12 < 15$$

Therefore the column is a short one.

$$\text{Gross area of concrete} = 250 \times 250 = 62\,500 \text{ mm}^2$$

$$A_{sc} = 4 \times 491 \text{ mm}^2 = 1964 \text{ mm}^2$$

$$A_c = A_{gross} - A_{sc} = 60\,536 \text{ mm}^2$$

Therefore

$$P_0 = p_{cc}A_c + p_{sc}A_{sc}$$
$$= 5.3 \times 60\,536 + 125 \times 1964$$
$$= 320.8 + 245.5$$
$$= 566.3 \text{ kN}$$

Example 12.8

A short reinforced concrete column is required to carry an axial load of 900 kN. Design a square column containing *a)* 0.8 per cent of steel and *b)* 8.0 per cent of steel.
   Assume $p_{cc} = 5.3 \text{ N/mm}^2$ and $p_{sc} = 125 \text{ N/mm}^2$.

Solution

$$a) \quad A_{sc} = \frac{0.8}{100} \times A_g$$

where A_g is gross cross-sectional area.
   Therefore

$$P_0 = p_{cc}A_c + p_{sc}A_{sc}$$
$$= 5.3(A_g - 0.008A_g) + 125 \times 0.008A_g$$
$$= 5.3A_g - 0.04A_g + 1.0A_g$$
$$= 6.26A_g$$

Hence

$$A_g = 900\,000/6.26 = 143\,770 \text{ mm}^2$$
$$\text{Length of side} = \sqrt{143\,770} = \text{(say)} \ 380 \text{ mm}$$
$$A_{sc} = 0.008 \times 0.144 \times 10^6 = 1155 \text{ mm}^2$$

Make column 380 mm square with 4 No. 20 mm dia. reinforcing bars (1260 mm^2).

b) $\quad A_{sc} = \dfrac{8.0}{100} \times A_g$

$\qquad P_0 = 5.3(A_g - 0.08A_g) + 125 \times 0.08A_g$

$\qquad \quad = 14.9A_g$

Hence

$\quad A_g = 9000\,000/14.9 = 60\,400$ mm^2

Length of side $\sqrt{60\,400} = $ (say) 250 mm

$A_{sc} = 0.08 \times 60\,400 = 4832$ mm^2

Make column 250 mm square with 4 No. 32 mm dia. and 4 No. 25 mm dia. reinforcing bars (5180 mm^2).

Example 12.9

A 300 mm square reinforced concrete column is axially loaded from four beams, one on each side. The column contains four 25 mm dia. bars and is 8 m high. Calculate the safe load if $p_{cc} = 5.3$ N/mm^2 and $p_{sc} = 125$ N/mm^2.

Solution Assuming partial fixity only, $l = 0.875L$, so

Slenderness ratio $= 0.875 \times 8000/300 = 23.3 > 15$

The column is slender and the appropriate reduction coefficient from Table 8 of CP114 is 0.71. Hence

Safe load $= 0.71 \times P_0$

where $\quad P_0 = 5.3(300^2 - 4 \times 491) + 125 \times 1960$

$\qquad \qquad = 466.6 + 245.5$

$\qquad \qquad = 712$ kN

The safe load for the slender column in question is

$0.71 \times 712 = 505.5$ kN

SUMMARY

Timber Columns. Permissible stresses for varying slenderness ratios are given in Table 12.1.

If the column is of solid rectangular cross-section, the slenderness ratio may be taken as: effective length divided by the least lateral dimension.

If not of rectangular cross-section, the slenderness ratio is: effective length divided by least radius of gyration.

The effective length should be assumed in accordance with Table 12.3. The safe axial load for the column is obtained by multiplying the permissible stress by the area of the cross-section of the column.

Steel Columns

$$\text{Slenderness ratio} = \frac{\text{effective length}}{\text{least radius of gyration}} = \frac{l}{r}$$

See Table 12.3 and Appendix D of BS449 for a guide in estimating effective lengths. Permissible stresses for varying slenderness ratios are given in Table 12.2.

The safe axial load is obtained by multiplying the permissible stress by the area of the cross-section of the column.

Least r for solid rectangular section $= 0.289b$

Least r for hollow square section $= \frac{1}{2}\sqrt{\left(\frac{B^2 + b^2}{3}\right)}$

Least r for hollow circular section $= \frac{1}{4}\sqrt{(D^2 + d^2)}$

Least r for rolled sections, e.g. $\mathbf{I}$ sections, channels and angles can be obtained from Tables in Structural Steelwork Handbooks or BS4.

Reinforced Concrete Columns. A short column is one where the ratio of the effective column length to least lateral dimension does not exceed 15. A large number of columns in buildings are, therefore, "short".

Safe axial load $P_0 = p_{cc}A_c + p_{sc}A_{sc}$

where p_{cc} = permissible concrete stress

A_c = area of cross-section of concrete

$\quad = A_g - A_{sc}$

p_{sc} = permissible steel stress

A_{sc} = area of steel.

EXERCISE 12

1 Calculate the safe axial loads for the following posts all of Grade S2/GS timber: *a*) 75 mm square, 2 m effective length; *b*) 100 mm square, 2.5 m effective length.

2 Calculate the safe axial loads for the following posts of Grade S2/GS timber all of 3 m effective length: *a*) 150 mm × 50 mm; *b*) 100 mm × 75 mm; *c*) 85 mm square.

3 A timber post of 3.6 m effective length is required to support an axial load of 230 kN. Determine the length of side of a square section post of Grade S2/SS timber.

4 A timber post of Grade S2/SS timber has an effective length of 3 m and is 300 mm diameter (solid circular cross-section). Calculate the value of the axial load.

5 A timber post of solid circular cross-section is of timber Grade S2/SS and 3 m effective length. It has to support an axial load of 300 kN. Determine the diameter of the post.

6 Two timber posts 200 mm × 75 mm in cross-section of Grade S2/GS timber are placed side by side without being connected together to form a post 200 mm × 150 mm. The effective length of the posts is 2.5 m. Calculate the safe axial load for the compound post and compare this load with the safe axial load of one solid post 200 mm × 150 mm in cross-section of the same effective length.

7 Calculate the safe axial load for a 254 × 254UC89 stanchion for the following end fixings and effective lengths:
a) 12 m, both ends fully restrained;
b) 9 m, one end fully restrained the other held in position;
c) 6 m, both ends held in position only.

8 Determine the maximum permissible effective length of a 254 × 254UC73 section. Calculate the safe axial load for such effective length.

9 Choose a suitable stanchion of ⊥ section to support an axial load of 980 kN, the effective length = 3 m.

10 Make calculations to determine which of the following stanchions is capable of supporting the greater axial load: *a*) A stanchion consisting of two 152 × 152UC23 sections placed side by side without being connected together; effective length = 2.6 m; *b*) One 203 × 203UC46 section of 2.6 m effective length.

11 Two 100 mm diameter solid round steel columns placed close together have an effective length of 4 m. Calculate the safe axial load for the columns. Calculate the diameter of one column which would be equivalent in load carrying capacity to the two columns.

12 A hollow round steel column is 75 mm outside diameter and the metal is 6 mm thick. Calculate the safe axial load for each of the effective lengths specified: 1.5 m, 2 m, 3 m.

13 Calculate the safe axial load for a 400 mm square "short" reinforced concrete column containing eight 25 mm diameter bars. Use 1 : 2 : 4 mix of concrete.

14 Calculate the safe axial load for a 450 mm diameter "short" circular cross-section reinforced concrete column containing six 32 mm diameter bars.

$$p_{cc} = 6 \text{ N/mm}^2 \quad p_{sc} = 125 \text{ N/mm}^2$$

15 Design a square cross-section column to support a load of 2 MN assuming 4 per cent of steel and $1 : 1\frac{1}{2} : 3$ concrete mix.

16 Design a short column to carry 2.5 MN, assuming:
a) $p_{cc} = 7.6 \text{ N/mm}^2$, $p_{sc} = 140 \text{ N/mm}^2$, 8 per cent steel.
b) $p_{cc} = 4.3 \text{ N/mm}^2$, $p_{sc} = 125 \text{ N/mm}^2$, 0.8 per cent steel.

17 Compare the load carrying capacities of a timber and a reinforced concrete column both 3 m effective length and 300 mm square. The timber is Group S2/SS and the reinforced concrete column contains eight 25 mm diameter bars.

$$p_{cc} = 6.0 \text{ N/mm}^2 \quad p_{sc} = 125 \text{ N/mm}^2$$

18 One dimension of a "short" reinforced concrete column must not exceed 250 mm. The column has to support an axial load of 730 kN. Assuming 4 per cent steel, design the column using 1:2:4 concrete mix.

19 The bottom length of a reinforced concrete column in a multi-storey building supports five floors and a flat roof. The area of each floor supported by the column is 36 m² and the *inclusive* loads are 4 kN/m² for the flat roof and 6 kN/m² for each floor. Design a suitable "short" column assuming 4 per cent steel.

$$p_{cc} = 6.5 \text{ N/mm}^2 \quad p_{sc} = 125 \text{ N/mm}^2$$

20 A column is 400 mm square and 9 m effective length and it contains eight 25 mm diameter bars. Calculate the safe axial load assuming

$$p_{cc} = 5.7 \text{ N/mm}^2 \quad \text{and} \quad p_{sc} = 125 \text{ N/mm}^2$$

21 A "short" column is 450 mm square and it has to support an axial load of 2 MN. Calculate the area of steel required.

$$p_{cc} = 6.0 \text{ N/mm}^2 \quad p_{sc} = 125 \text{ N/mm}^2$$

22 A square concrete column with 5 per cent of reinforcement is to carry an axial load of 554 kN.
Given that $p_{cc} = 8 \text{ N/mm}^2$ and $p_{sc} = 125 \text{ N/mm}^2$ determine:
a) a suitable size for the column and the cross-sectional area of the reinforcement;
b) the maximum length of the column so that it may be treated as a "short" column, assuming full restraint at both ends.

13 Connections

The treatment of connections in this chapter is confined to direct shear connections in steelwork. The behaviour of rivets and bolts is presented in some detail. The criterion value of rivets or bolts is explained and its use in design calculations is demonstrated. High Strength Friction Grip Bolts are dealt with briefly to give an introduction to this type of connection. Finally, welding is also discussed because of its undoubted versatility and the resulting popularity.

RIVETING AND BOLTING

A rivet or bolt may be considered simply as a *peg* inserted in holes drilled in two or more thicknesses of steel in order to prevent relative movement. For example, the two steel plates in Fig. 13.1 tend to slide over each other, but could be prevented from doing so by a suitable steel pin inserted in the holes in each plate, as shown. In order to prevent the *steel pin* from slipping out of holes, bolts with heads and nuts are used or rivet heads are formed, and these produce an effective connection (Fig. 13.2).

The rivet heads (or bolt heads and nuts) do, in fact, strengthen the connection by pressing the two thicknesses of plate together, but this strength cannot be determined easily, and so the rivet or bolt strength is calculated on the assumption that its shank (shown shaded) only is used in building up its strength.

Figure 13.1

Figure 13.2

Single Shear

If the loads W in Fig. 13.2 are large enough, the rivet or bolt could fail, as in Fig. 13.3, in *shear*, i.e. breaking by the sliding

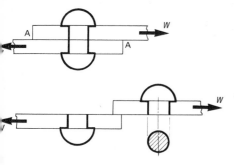

Figure 13.3

of its fibres along line A–A. This type of rivet or bolt failure is known as **failure in single shear.** The area of steel rivet resisting this failure is the circular area of the rivet shank, shown hatched in Fig. 13.3, i.e.

$$\tfrac{1}{4}\pi \times (\text{diameter of rivet})^2 \quad \text{or} \quad 0.7854d^2 = A$$

The permissible shear stress for bolts of strength grade 4.6 and mild steel rivets in single shear is given in BS449 as follows:

a) Black bolts
b) Hand-driven rivets $\Big\}$ 80 N/mm²
c) Power-driven rivets
d) Close tolerance and $\Big\}$ 100 N/mm²
   turned bolts

Black bolts are manufactured from rolled steel bars and, owing to the difficulties in rolling round black bars to the exact required diameter, it is customary to drill the holes in the plates, etc., 2 mm greater in diameter than the specified diameter of the bolt.

Close tolerance and turned bolts are made from rolled steel bars which are greater in diameter than the required size of the bolt. The bars are then turned down to the required diameter. These bolts give a better fit in the holes than black bolts and are therefore allowed higher stresses. Power-driven rivets are usually driven by a special machine. The rivets and the rivet heads are formed more accurately than is possible in the case of hand-driven rivets and they are therefore permitted a higher stress. The holes are drilled 2 mm larger in diameter than the specified sizes of the rivets.

Since rivets are driven while hot and, therefore, their material fills the hole completely, it is necessary to distinguish between the nominal and the gross diameter of the rivets. The nominal diameter refers to the specified size of the rivet shank, i.e. the diameter of the rivet when it is cold, whilst the gross diameter is 2 mm larger than the specified (i.e. nominal) diameter of the rivet. BS449 allows the strength of a rivet to be estimated on its *gross* diameter.

For example, the safe load in single shear (safe stress × area) of a 16 mm diameter power-driven rivet is

$$100 \times 0.7854 \times 18^2 = 25.4 \text{ kN}$$

For bolts, the gross diameter is, of course, equal to the nominal diameter. Therefore the safe load in single shear, or single shear value (s.s.v) of a 16 mm diameter black bolt of strength grade 4.6, is

$$80 \times 0.7854 \times 16^2 = 16 \text{ kN}$$

Figure 13.4

Double Shear

In the type of connection shown, for example, in Fig. 13.4 (a double cover butt joint), the rivets or bolts on one side of the joint would have to shear across *two* planes, as shown. This is known as **failure in double shear.**

A rivet or bolt under these circumstances will need twice as much load to break it compared with a rivet or bolt in single shear, so the safe load on a rivet in double shear is twice that on the same rivet in single shear. Thus, using the circular area of the rivet shank as before,

Safe load in double shear
$$= \text{area} \times 2 \times \text{single shear safe stress}$$

Therefore,

Safe load for the above black bolt
$$= 2 \times \text{permissible shear stress} \times \text{area}$$
$$= 2 \times 80 \times 201$$
$$= 32 \text{ kN}$$

or, simply,

$$2 \times \text{s.s.v.} = 2 \times 16 = 32 \text{ kN}$$

Bearing

The two main ways in which the rivet or bolt itself may fail have been discussed. This type of failure assumes, however, fairly thick steel plates capable of generating sufficient stress to shear the rivet.

Consider Fig. 13.5(*a*). The heavy load of 120 kN taken by the 25 mm steel plates would certainly shear the 12 mm diameter rivet (single shear).

Now consider the opposite type of case, as in Fig. 13.5(*b*), where a thick steel rivet (24 mm diameter) is seen connecting two very thin steel plates. The steel plates in this case are much more likely to be torn by the rivet than the rivet to be sheared by the weaker steel plates.

This type is known as **failure in bearing** (or *tearing*), and note should again be taken of the area which is effective in

Figure 13.5

Section A-A

Figure 13.6

Area resisting bearing = 12 x 22mm²

Figure 13.7

Area resisting bearing = 10 x 22mm

Figure 13.8

Minimum area resisting
bearing 25 x 18 = 450mm

Figure 13.9

resisting this type of stress (Fig. 13.6). The area of contact of the rivet with the plate on one side of it is actually semi-cylindrical, but since the bearing stress is not uniform, it is assumed that the area of contact is the thickness of plate times diameter of rivet. This area is shown shaded in section A–A of Fig. 13.6

For bearing purposes as for shear, the gross diameter of the rivet can be taken as the nominal diameter plus 2 mm.

When two plates of the same thickness are being connected, then of course *either* plate could tear, and the area resisting bearing would be the thickness of one plate times diameter of rivet (Fig. 13.7). Where plates of different thicknesses are used, then the thinner of the two plates would tear first, so the area resisting bearing or tearing would be the thickness of thinner plate times the diameter of rivet (Fig. 13.8). Where three thicknesses are concerned, as in Fig. 13.9, the two 15 mm plates are acting together and the 25 mm plate would tear before the two 15 mm plates, so the area resisting tearing would be $25 \times 18 = 450$ mm².

The permissible stress in bearing for bolts of strength grade 4.6 and mild steel rivets is given in BS449 as follows:

a) Black bolts
b) Hand-driven rivets } 250 N/mm²
c) Power-driven rivets }
d) Close tolerance and } 300 N/mm²
   turned bolts

Criterion Value

It will be seen that rivets or bolts may be designed on the basis of a) their strength in shear, or b) their strength in bearing.

In actual design, the lesser of these two values will, of course, have to be used. This is called the **criterion value** of the rivet or bolt.

When designing this type of connection, the following questions should be asked:

1 Is the connection in single or double shear?
2 What is the safe appropriate shear load on one rivet or bolt?
3 What is the safe bearing load on one rivet or bolt?

The criterion value, that is the lower of the values given by (2) or (3), will then be used in determining the safe load on the connection.

Example 13.1

Four 20 mm diameter power driven rivets

Figure 13.10

Calculate the safe load W on the lap joint shown in Fig. 13.10 in terms of the shear or bearing values.

Solution There are only two plates, and therefore the connection is in single shear.
  Single shear value of one rivet is

$$100 \times 0.7854 \times 22^2 = 100 \times 380 = 38 \text{ kN}$$

Bearing value of the rivet in 12 mm plate is

$$300 \times 22 \times 12 = 300 \times 264 = 79 \text{ kN}$$

The shear value is, therefore, the criterion value in this case, and the safe load for the connection is

$$4 \times 38 = 152 \text{ kN}$$

Example 13.2

What is the safe load W in kN on the tie shown in Fig. 13.11 with respect to rivets A and B?

Solution Rivets A are in double shear and rivets B are in single shear.

Rivets A
D.S.V. of one 20 mm diameter rivet is

$$2 \times 100 \times 380 = 76 \text{ kN}$$

Bearing value of that rivet in 15 mm plate is

$$300 \times 22 \times 15 = 99 \text{ kN}$$

Criterion value $= 76 \text{ kN}$

Therefore, total value of rivets A is

$$3 \times 76 = 228 \text{ kN}$$

Rivets B
S.S.V. of one 20 mm diameter rivet is

Figure 13.11

$$100 \times 380 = 38 \text{ kN}$$

Bearing value of that rivet in 12 mm plate is

$$300 \times 22 \times 12 = 79 \text{ kN}$$

Criterion value $= 38$ kN

Therefore, total value of rivets B is

$$4 \times 38 = 152 \text{ kN}$$

It appears, therefore, that the safe load W is decided by rivets B and is 152 kN. The strength of the plate in tension, however, should be investigated before the 152 kN load is accepted as the safe load.

The 125 mm $\times$ 12 mm plate is weakened by one 22 mm diameter rivet hole. Hence the net cross-sectional area of the plate is

$$(125 - 22) \times 12 = 1236 \text{ mm}^2$$

and the permissible tensile stress for Grade 43 steel is 155 N/mm² for plates up to 40 mm thick.

Tension value of the plate is

$$155 \times 1236 = 191 \text{ kN}$$

In this case, therefore, the strength of the connection is decided by rivets B. It should be noted, however, that the angles have not been checked, although they are usually satisfactory.

Example 13.3

A compound bracket, shown in Fig. 13.12, is connected to the 13 mm thick web of a stanchion by six 16 mm diameter close tolerance bolts of strength grade 4.6. The bracket carries a reaction of 150 kN from a beam. Is the connection strong enough in terms of the bolts?

Solution There are three thicknesses—the web, the angle and the cover plate—but the bolts are in single shear because the angle and the cover plate act as one.

S.S.V. of one 16 mm diameter bolt is

$$100 \times 0.7854 \times 16^2 = 100 \times 201 = 20.1 \text{ kN}$$

Bearing value of the bolt in 13 mm plate is

$$300 \times 16 \times 13 = 300 \times 208 = 62.4 \text{ kN}$$

Criterion value = Shear value = 20.1 kN

Safe load $= 6 \times 20.1 = 120.6$ kN

Small connecting angle is ignored in calculations

457 x 152 UB60

A

12mm thick L

12mm plate

Six 16mm close tolerance bolts

A

Web = 13mm thick

Figure 13.12

This is less than the applied load (reaction). Therefore, either the number of the 16 mm diameter bolts should be increased or larger diameter bolts will have to be used.

In practice the bolts in this type of connection would be also investigated for direct tension since, according to clause 34.a. of BS449 : Part 2 : 1969, the reaction must be assumed to be applied at least 100 mm from line A–A (Fig. 13.12), thus creating an eccentricity of loading.

Double Cover Butt Connections

In designing butt and other similar types of connections, it should always be borne in mind that not only can failure occur through an insufficient number of rivets or bolts being provided, but that the member itself may fail in tension.

Consider, for example, Fig. 13.13 noting, in particular, the lay-out of the rivets in what is called a "leading rivet" arrangement.

One possible chance of failure is that the plate being connected would fail by tearing across face A–A or B–B under a heavy load. Therefore, no matter how many rivets are employed, the safe strength in tension across this and other faces could never be exceeded.

The strength of the rivets must be approximately equal to the strength of the member in tension for the connection to be considered economical.

Figure 13.13

Example 13.4

A 150 mm × 18 mm steel plate used as a tension member in a structural frame has to be connected using a double cover butt connection with two 12 mm cover plates and 20 mm diameter power-driven rivets.

Design a suitable connection assuming that the permissible stress in tension for the steel plate is 155 N/mm².

Solution However the rivets are arranged, the section will be weakened by having at least one rivet hole so the net cross-sectional area of the plate is

$$(150 - 22) \times 18 = 2304 \text{ mm}^2$$

and the safe load carried by the plate must not exceed

$$155 \times 2304 = 357 \text{ kN}$$

The rivets will be in double shear.
D.S.V. of one 20 mm diameter rivet is

$$2 \times 100 \times 380 = 76 \text{ kN}$$

357 kN ⟵   ⟶ 357 kN

18 mm plate

12 mm cover plates

Figure 13.14

Bearing value of that rivet in 18 mm plate is

$$300 \times 22 \times 18 = 118.8 \text{ kN}$$

Criterion value = 76 kN

Therefore, number of rivets required on each side of joint is

$$\frac{\text{Total load}}{\text{Value of one rivet}} = \frac{357}{76} = 4.7, \text{ say 5 rivets}$$

The arrangement of the rivets is shown in Fig. 13.14.

Check the strength of the plate:
At section A–A, the strength, as calculated, is 357 kN.

At section B–B, the plate is weakened by two rivet holes, but, in the event of tearing of the plate across B–B, the connection would not fail until the rivet marked x also had failed.

Thus the strength across B–B is

$$155 \times (150 - 2 \times 22) \times 18 + 76$$
$$= 155 \times 1908 + 76 = 371.7 \text{ kN}$$

and the strength at C–C is

$$155 \times 1908 + 3 \times 76 = 523.7 \text{ kN}$$

since in this case the rivets x, y and z have to be considered.

Finally, the cover plates have to be designed. The critical section here is section C–C. Assume 12 mm thick plates then the strength at C–C is

$$155 \times (150 - 2 \times 22) \times 24 = 394 \text{ kN}$$

Therefore the above connection would carry 357 kN.

It is sometimes useful to check the "efficiency" of the connection. This is given by

$$\text{Efficiency} = \frac{\text{Safe load for the connection}}{\text{Original value of the undrilled plate}} \times 100$$

In the above case, the efficiency would be

$$\frac{357\,000}{155 \times 150 \times 18} \times 100 = 85.3\%$$

Increasing the number of rivets above that which is required may, in some cases, actually weaken the connection.

Consider the connection in Example 13.4. Had six rivets been used as in Fig. 13.15, instead of the required five, the value of the plate at section A–A would now be

$$155 \times (150 - 2 \times 22) \times 18 = 296 \text{ kN}$$

as against 357 kN for the leading rivet arrangement.

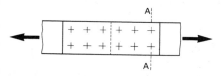

Figure 13.15

HIGH STRENGTH FRICTION GRIP BOLTS

The rivets and bolts, discussed so far in this chapter, relied on their shear and bearing strength to produce an effective connection capable of transmitting a load from one member to another, e.g. from beam to column.

The performance of High Strength Friction Grip (HSFG) bolts is based on the principle that the transfer of the load may be effected by means of friction between the contact surfaces (interfaces) of the two members. To produce the necessary friction a sufficiently high clamping force must be developed, and this is achieved by tightening the bolts to a predetermined tension. In this way the bolts are subjected to a direct (axial) tensile force and do not rely on their shear and bearing strength.

The substantial forces needed to produce the necessary "friction grip" require the bolts to be of high tensile strength and the interfaces to be meticulously prepared. Various methods are used to ensure that the bolts are tightened to the required tension in the shank. These include torque control by means of calibrated torque wrenches and special load indicating devices.

HSFG bolts and their use are specified in BS4395 and BS4604, respectively, and further details may be obtained from manufacturers' literature.

Considering connections subject only to shear between the friction faces, the safe load may be determined from the following:

$$\frac{\text{Slip factor}}{\text{Load factor}} \times \text{number of effective interfaces}$$

$$\times \text{proof load of one bolt}$$

$$\times \text{number of bolts}$$

Table 13.1 High Strength Friction Grip Bolts (from BS4395:Part 1:General Grade)

Nominal size and thread diameter	Proof load (minimum shank tension) kN
M12	49.4
M16	92.1
M20	144
M24	207
M27	234
M30	286
M36	418

The slip factor is really the coefficient of friction between the surfaces and may be taken as 0.45 for surfaces complying with the appropriate specification.

The load factor for structures covered by BS449 is usually at least 1.4, although, in cases where wind forces are considered, it may be reduced to 1.2.

Effective interface is the common contact surface, i.e. a single shear connection has one effective interface and a double shear connection has two.

Proof load is the minimum shank tension depending on the size of the bolt (See Table 13.1).

Example 13.5

Consider the connection in Example 13.3 (Fig. 13.12). Assume that the six bolts are 16 mm diameter HSFG bolts (general grade). Is the connection strong enough now?

Solution

$$\text{Safe load} = \frac{0.45}{1.4} \times 1 \times 92.1 \times 6 = 177 \text{ kN} > 150 \text{ kN}$$

i.e. the connection is now satisfactory.

It must be pointed out again that here, as in the case of Example 13.3, the bolts would also be subject to tension caused by the eccentricity of loading. This tension reduces the effective clamping action of the bolts and, therefore, the safe load would have to be suitably decreased.

WELDING

Welding for structural purposes is governed by the requirements of BS5135 *Metal-arc welding of carbon and carbon-manganese steels* and the design of welds is covered by clause 54 of BS449.

The two types of weld used are butt welds and fillet welds.

1 **Butt welds** These require the edges of the plates to be prepared by bevelling or gouging as shown in Fig. 13.16. This preparation and the need for careful alignment when welding make the butt weld generally the more expensive of the two.

For the purpose of strength calculations, butt welds are treated as the parent metal, i.e. the allowable stresses for the weld are the same as those for the connected plates.

Single V   Single U   Single J   Single Bevel

(a)   Butt welds

Leg Length

Throat thickness

(b)   Fillet welds

Figure 13.16

In (*b*), throat thickness
= leg length × sin 45°
= 0.7 × leg length

2 **Fillet welds** No special preparations are needed and the strength of the weld is calculated on the throat thickness (see Fig. 13.16).

The allowable stress depends on the grade of the steel of the connected parts and is 115 N/mm² for grade 43 steel and 160 N/mm² for grade 50 steel.

The size of the weld is specified by the minimum leg length of the weld, e.g. the strength of a 8 mm fillet weld for grade 43 steel is

$$8 \times 0.7 \times 115 = 644 \text{ N/mm}$$

i.e. each millimetre length of this weld is capable of carrying a load of 644 N.

When deciding on the size of a weld it is well to consider that the amount of weld metal increases faster than the strength of the weld, e.g. compare 6 mm and 8 mm welds:

increase in strength 33%
increase in weld metal 78%

Example 13.6

A tension member in a framework consists of an 80 mm × 10 mm flat and is subject to a direct force of 110 kN. Design a suitable fillet weld connection using a gusset plate, as shown in Fig. 13.17.

Solution Welding along the two edges of the flat requires a minimum length of weld of 80 mm on each side (Cl. 54.f. BS449), i.e. minimum length of weld is 160 mm.

Use 6 mm weld

$$\text{Required length} = \frac{110\,000}{6 \times 0.7 \times 115} = 228 \text{ mm}$$

The weld should be returned continuously around the corner for a distance not less than 2×weld size to comply with clause 54.e. of BS449 and an allowance of one weld size should be made at the open end of the weld.

The overall length of the welds should be

$$\tfrac{1}{2} \times 228 + l \times 6 + \text{return end}$$

6 mm
fillet
weld

Gusset plate

120 mm

80 mm
×10 mm

110 kN

Figure 13.17

SUMMARY

Rivets and Bolts

S.S.V. of one rivet or bolt $= Af_q$

D.S.V. of one rivet or bolt $= 2Af_q$

B.V. of one rivet or bolt in a plate of thickness t mm $= dtf_b$

A is the area of cross-section of the rivet shank or bolt shank.

For rivets, A may be taken as the area of a circle 2 mm greater in diameter than the specified (nominal) diameter.

For bolts, A is the area calculated from the nominal diameter.

d is the diameter of the rivet or bolt.

For rivets, d = nominal diameter plus 2 mm

For bolts, d = nominal diameter

f_q = permissible shear stress

f_b = permissible bearing stress

In certain problems, the strength of the plate in tension may have to be investigated. The permissible tension stress for grade 43 steel is 155 N/mm².

When deducting the areas of rivet or bolt holes to determine the strength of a plate, the diameter of the hole is taken as 2 mm greater than the nominal diameter of the rivet or bolt.

HSFG bolts rely on their tensile strength to induce friction between the connected parts.

A **butt weld** is considered to be as strong as the parent metal.

The strength of a **fillet weld** per millimetre of its length is calculated as

$$0.7 \times \text{size of weld} \times \text{allowable stress}$$

Figure 13.18

EXERCISE 13

(*Note:* permissible tension stress for Grade 43 steel = 155 N/mm²)

1 The size of each plate in a simple lap joint is 100 mm × 12 mm and there are six 20 mm diameter turned bolts in a single line. Calculate the safe load in tension.

2 In a double cover butt connection, the joined plate is 125 mm × 12 mm and the cover plates are 125 mm × 8 mm. There are two 20 mm diameter power-driven rivets each side of the joint (four rivets in all). Calculate the maximum safe tension for the plates.

3 A simple lap joint with five 24 mm diameter hand-driven rivets is shown in Fig. 13.18. Calculate the maximum safe pull, W.

Figure 13.19

Figure 13.20

Figure 13.21

4 Eight black bolts, 24 mm diameter, connect a flat bar to a gusset plate as shown in Fig. 13.19. Calculate the safe load W.

5 Fig. 13.20 gives two different bolted connections (*a*) and (*b*). In each case, calculate the safe load W.

6 A 100 mm wide plate is connected, by means of a 10 mm gusset plate to the flange of a 254×254UC107 stanchion, as shown in Fig. 13.21. Calculate the required thickness of the plate and the number of 20 mm diameter black bolts.

7 The end of a beam is supported by a hanger consisting of a 80 mm×80 mm×8 mm angle section, as shown in Fig. 13.22. The load at the end of the beam is 100 kN. Calculate the number of 20 mm diameter turned bolts required along the line AB and check the strength in tension of the angle section. (Gross area of angle = 1230 mm^2.)

8 A 10 mm thick tie member is to be connected to a 10 mm thick gusset plate, as indicated in Fig. 13.23. Calculate the necessary width x of the tie and calculate the number of 16 mm diameter black bolts to connect the tie to the gusset.

9 Fig. 13.24 shows a joint in a tension member. Determine the safe load W. (Calculations are required for the strength of the middle plate at sections A–A and B–B; the strength of the cover plate at C–C, and the strengths of the rivets in shear and bearing.)

Figure 13.22

Figure 13.23

Figure 13.24

All rivets 16 mm dia. hand driven rivets

Figure 13.25

150 mm x 90 mm x 10 mm L

10 mm plate

100 kN

Figure 13.26

Figure 13.27

10 Referring to Fig. 13.25, calculate the maximum safe load W, confining the calculations to that part of the connection which lies to the right of line A–A. Then, using this load, find the required thickness of the 75 mm wide plate at 140 N/mm^2 and the necessary number of 16 mm diameter hand-driven rivets.

11 It is proposed to suspend a load of 100 kN from an existing beam in a building by the means shown in Fig. 13.26, using 20 mm diameter turned bolts. Calculate the width of plate required and the number of bolts.

12 Calculate the value of the maximum permissible load W for the connection shown in Fig. 13.27. The rivets are 20 mm diameter, hand-driven.

The above examples may be adapted for HSFG bolts and fillet welds.

14 Addition of Direct and Bending Stress

Previous chapters have dealt largely with two main types of stress:

1 *Direct or Axial Stress* This occurs when a load (tensile or compressive) is spread evenly across a section, as in the case of a short column axially loaded. Here, the unit stress or stress per mm² is found by dividing the total load by the sectional area of the member, i.e.

Direct stress $= f_d = W/A$

2 *Bending Stress* In this case, the stress has been seen to vary—having different values at different distances from the neutral axis of the section. The stress at a distance y from the neutral axis is $f_b = M \times y/I$, or where the stress at the extreme fibres is required $f_{b\,max} = M/Z$.

Cases frequently arise, however, where *both* these types of stress occur at the same time, and the combined stress due to the addition of the two stresses is to be determined.

Consider, for example, the short timber post shown in plan in Fig. 14.1. The load passes through the centroid of the section, and the resulting stress is pure axial stress

$W/A = 90\,000/22\,500 = 4\ \text{N/mm}^2$

The stress is the same at all points of the section, and this is shown by the rectangular shape of the "stress distribution diagram."

Fig. 14.2(*a*) shows the same section with a load of 90 kN as before, but this time the load, whilst still on the X–X axis, lies 20 mm away from the Y–Y axis.

There will still be a direct stress everywhere on the section as before of $W/A = 4\ \text{N/mm}^2$, but the eccentricity of the load with regard to the Y-Y axis causes a moment of $90\,000 \times 20 = 1.8 \times 10^6\ \text{N mm}$, which has the effect of increasing the compression on the area to the right of Y–Y (shaded) and

(a)

$\dfrac{W}{A} = \dfrac{90000}{150 \times 150} = 4\,\text{N/mm}^2$

(b)

Figure 14.1 (*a*) Loading; (*b*) Stress distribution

Figure 14.2

Figure 14.3

decreasing the direct compression on the portion to the left of Y–Y.

Fig. 14.2(b) shows the direct stress W/A in compression. Fig. 14.2(c) shows the bending stress of

$$\frac{M}{Z} = \frac{1\,800\,000 \times 6}{150^3} = 3.2 \text{ N/mm}^2 \text{ compression at BC}$$

$$\text{and} \quad 3.2 \text{ N/mm}^2 \text{ tension at AD}$$

Fig. 14.2(d) shows how these two stress distribution diagrams (b, c) are superimposed, giving the final diagram as shown. The compressive stress on the edge BC is

$$\frac{W}{A} + \frac{M}{Z} = 4.0 + 3.2 = 7.2 \text{ N/mm}^2$$

whilst the stress on edge AD is

$$\frac{W}{A} - \frac{M}{Z} = 4.0 - 3.2 = 0.8 \text{ N/mm}^2$$

Note: The stress is still compressive at any point, but it varies from face AD to face BC.

The stress at section Z–Z (25 mm from Y–Y) is

$$\frac{W}{A} + \frac{My}{I_{yy}} \quad \text{where } y = 25 \text{ mm}$$

$$= 4.0 + \frac{1\,800\,000 \times 25 \times 12}{150^4}$$

$$= 4.0 + 1.1 = 5.1 \text{ N/mm}^2$$

Fig. 14.3(a) shows the same short column as in the previous example, but this time the eccentricity about the Y–Y axis has been increased to 25 mm. As before,

Direct stress $= W/A = 90\,000/22\,500 = 4 \text{ N/mm}^2$

but the M is now $90\,000 \times 25 = 2.25 \times 10^6$ N mm. Thus

$$\frac{M}{Z} = \frac{2.25 \times 10^6}{562\,500} = 4 \text{ N/mm}^2$$

Figs. 14.3(b) and 14.3(c) show the direct and bending stresses respectively, whilst in Fig. 14.3(d) the two diagrams have been superimposed, showing that

1) The stress on face BC is

$$\frac{W}{A} + \frac{M}{Z} = 4.0 + 4.0 = 8.0 \text{ N/mm}^2$$

2) The stress on face AD is

$$\frac{W}{A} - \frac{M}{Z} = 4.0 - 4.0 = 0$$

The original compression of W/A on face AD has been cancelled out by the tension at AD caused by the eccentricity of the loading away from that face, and the stress intensity at AD $= 0$, whilst that on face BC $= 8.0\,\text{N/mm}^2$.

This interesting example demonstrates a general rule that, in the case of square or rectangular sections, the load may be eccentric from one axis by no more than $\frac{1}{6}$th of the width unless tension on one face is permissible.

In the case of a steel, timber or reinforced concrete column tension would be allowable within reasonable limits, but, in a brick wall, tension would be undesirable and the resultant load or thrust in a masonry structure should not depart more than $d/6$ (or $b/6$) from the axis.

This is usually called the **Law of the Middle Third,** and it states that if the load or the resultant thrust cuts a *rectangular* section at a point not further than $d/6$ from the centre line (i.e. within the middle third), then the stress everywhere across the section will be compressive.

Consider the masonry retaining wall shown in Fig. 14.4(a). The earth pressure and the weight W of the wall combine to form a resultant R which cuts the base, as shown, at a distance e from the neutral axis (which, in this case, is also the centre line) of the base.

The vertical component V of the resultant causes a moment of $V \times e$, which adds compressive stress between the wall and the earth below at A, and reduces the pressure between the surfaces at B.

If V cuts the base within the middle third of its width d, then the stress everywhere between A and B will be compressive, as in Fig. 14.4(b).

Note: As the wall itself may be of any length, it is customary in design to consider only one metre length of wall, i.e. to estimate the weight of one metre of wall and the earth pressure on one metre of length also.

Thus, if the rectangle shown in Fig. 14.4(c) represents the area of wall resting on the base AB (as seen in a plan), then the eccentricity of the load V will tend to rotate this area about the axis Y–Y, so that, in calculating Z, the value $bd^2/6$ will be in effect $1 \times d^2/6 = \frac{1}{6}$th of d^2, and will be normally in m^3 units.

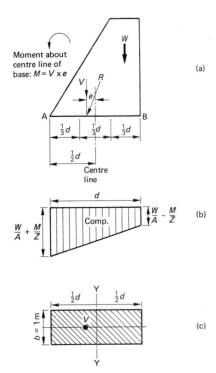

Figure 14.4

Eccentricity about Both Axes

The short pier shown (Fig. 14.5) carries one load only, eccentric to both X–X and Y–Y axes. Consider the stresses at A, B, C and D.

At C, the stress is

Figure 14.5

$$\frac{W}{A} + \frac{M_{xx}}{Z_{xx}} + \frac{M_{yy}}{Z_{yy}}$$

because the load has moved towards C from both axes. Stress at corner B will be

$$\frac{W}{A} - \frac{M_{xx}}{Z_{xx}} + \frac{M_{yy}}{Z_{yy}}$$

for the load is away from the Y–Y axis on the same side as B, but is away from the X–X axis on the side furthest from B. Similarly,

$$\text{Stress at corner A} = \frac{W}{A} - \frac{M_{xx}}{Z_{xx}} - \frac{M_{yy}}{Z_{yy}}$$

$$\text{Stress at corner D} = \frac{W}{A} + \frac{M_{xx}}{Z_{xx}} - \frac{M_{yy}}{Z_{yy}}$$

The solution is outlined below:

$$\frac{W}{A} = \frac{450\,000}{300 \times 500} = 3.0 \text{ N/mm}^2$$

$$\frac{M_{xx}}{Z_{xx}} = \frac{450\,000 \times 50 \times 6}{300 \times 500^2} = 1.8 \text{ N/mm}^2$$

$$\frac{M_{yy}}{Z_{yy}} = \frac{450\,000 \times 25 \times 6}{500 \times 300^2} = 1.5 \text{ N/mm}^2$$

Hence,

Stress at corner A $= 3.0 - 1.8 - 1.5 = -0.3 \text{ N/mm}^2$ tension

Stress at corner B $= 3.0 - 1.8 + 1.5$
$$= +2.7 \text{ N/mm}^2 \text{ compression}$$

Stress at corner C $= 3.0 + 1.8 + 1.5$
$$= +6.3 \text{ N/mm}^2 \text{ compression}$$

Stress at corner D $= 3.0 + 1.8 - 1.5$
$$= +3.3 \text{ N/mm}^2 \text{ compression}$$

Example 14.1

A short stanchion carries three loads, as shown in Fig. 14.6. Calculate the intensity of stress at corners A, B, C and D.

$$\text{Area} = 7580 \text{ mm}^2 \quad Z_{xx} = 580\,000 \text{ mm}^3 \quad Z_{yy} = 200\,000 \text{ mm}^3$$

Solution Method 1

$$\text{Direct stress } \frac{W}{A} = \frac{400 + 120 + 100}{7580} = 81.8 \text{ N/mm}^2$$

$$\text{Bending stress } \frac{M_{xx}}{Z_{xx}} = \frac{120\,000 \times 205}{580\,000} = 42.4 \text{ N/mm}^2$$

203 x 203 UC60

Figure 14.6

Bending stress $\dfrac{M_{yy}}{Z_{yy}} = \dfrac{100\,000 \times 20}{200\,000} = 10.0\,\text{N/mm}^2$

Therefore, the combined stresses are:

at A $81.8 - 42.4 - 10.0 = +\ 29.4\,\text{N/mm}^2$

at B $81.8 + 42.4 - 10.0 = +114.2\,\text{N/mm}^2$

at C $81.8 + 42.4 + 10.0 = +134.2\,\text{N/mm}^2$

at D $81.8 - 42.4 + 10.0 = +\ 49.4\,\text{N/mm}^2$

Method 2

The value and position of the resultant of the three loads is calculated and the resultant load applied to the column in place of the three loads.

To find distance x (see Fig. 14.7) take moments of loads about the X–X axis:

$(400 + 120 + 100) \times x = 120 \times 205$

$x = 24\,600/620 = 39.7\,\text{mm}$

For distance y take moments about the Y–Y axis:

$620 \times y = 100 \times 20$

$y = 2000/620 = 3.22\,\text{mm}$

Now, the direct stress W/A remains the same and the bending stresses are

$$\frac{M_{xx}}{Z_{xx}} = \frac{620\,000 \times 39.7}{580\,000} = 42.4\,\text{N/mm}^2$$

$$\frac{M_{yy}}{Z_{yy}} = \frac{620\,000 \times 3.22}{200\,000} = 10.0\,\text{N/mm}^2$$

Both are the same as in Method 1.

Application to Prestressed Concrete Beams

The addition of direct and bending stresses is not confined to columns only. It is used extensively in the design calculations for prestressed concrete beams. The prestressing force is generally applied in the direction of the longitudinal axis of the beam, thus producing direct stresses in the beam. The bending stresses result mainly from the loads the beam has to support, but there is usually also bending caused by an eccentricity of the prestressing force.

Example 14.2

A plain concrete beam, spanning 6 m, is to be prestressed by a force of 960 kN applied axially to the 240 mm × 640 mm

Figure 14.7

beam section.

Calculate the stresses in the top and bottom fibres of the beam if the total inclusive udl carried by the beam is 136 kN.

Solution The direct stress caused by the prestressing force P is

$$\frac{P}{A} = \frac{960\,000}{240 \times 640} = 6.25 \text{ N/mm}^2$$

The bending stresses, on the other hand, are produced by the inclusive udl and these are M/Z where

$$M = \tfrac{1}{8}Wl = \tfrac{1}{8} \times 136 \times 6 \times 10^6 = 102 \times 10^6 \text{ N mm}$$

$$Z = \tfrac{1}{6}bd^2 = \tfrac{1}{6} \times 240 \times 640^2 = 16\,384\,000 \text{ mm}^3.$$

Therefore,

$$\frac{M}{Z} = \frac{102\,000\,000}{16\,384\,000} = 6.23 \text{ N/mm}^2$$

and the combined stresses are

at the top $\quad 6.25 + 6.23 = +12.48 \text{ N/mm}^2$

at the bottom $\quad 6.25 - 6.23 = +0.02 \text{ N/mm}^2$

as shown diagrammatically in Fig. 14.8.

Figure 14.8 (*a*) Beam details; (*b*) Stress diagrams

In order to increase the beam's load bearing capacity it is usual to apply the prestressing force below the centre line of the section. This induces tensile bending stresses in the top fibres and compressive bending stresses in the bottom fibres of the beam. The applied loads produce bending stresses of the opposite sign, i.e. tension in bottom and compression in top fibres.

The combination of the stresses is shown in Fig. 14.9.

Figure 14.9 (*a*) Beam detail; (*b*) Stress diagrams

Example 14.3

Consider the beam in Example 14.2 with the prestressing force applied on the vertical axis of the section but 120 mm below its intersection with the horizontal axis.

Calculate the stresses in top and bottom fibres.

Solution

Direct stress P/A (as before) $\qquad = +6.25 \text{ N/mm}^2$

Bending stress due to prestressing force $\dfrac{P \times e}{Z} = \dfrac{960\,000 \times 120}{16\,384\,000} = \mp 7.03 \text{ N/mm}^2$

Bending stress due to applied loads $\dfrac{M}{Z} = (\text{as before}) \qquad \pm 6.23 \text{ N/mm}^2$

The combined stresses are

at the top $\quad +6.25 - 7.03 + 6.23 = +5.45 \text{ N/mm}^2$

at the bottom $\quad +6.25 + 7.03 - 6.23 = +7.05 \text{ N/mm}^2$

It may be concluded from this Example that the udl could be doubled without indicating tensile stresses in the concrete when the beam is fully loaded. There are, however, various problems which have to be taken into account when designing prestressed concrete beams and which are outside the scope of this textbook.

SUMMARY

Maximum combined stress = direct stress + bending stress. When load W is eccentric to axis X–X only:

$$\text{Maximum stress} = \frac{W}{A} + \frac{We_x}{Z_{xx}}$$

When load W is eccentric to axis Y–Y only:

$$\text{Maximum stress} = \frac{W}{A} + \frac{We_y}{Z_{yy}}$$

When load W is eccentric to both axes:

$$\text{Maximum stress} = \frac{W}{A} + \frac{We_x}{Z_{xx}} + \frac{We_y}{Z_{yy}}$$

In prestressed concrete beams, where opposing bending stresses are produced by the prestressing force P and the applied loads W,

$$\text{Maximum stress} = \frac{P}{A} \mp \frac{P \times e}{Z} \pm \frac{M_w}{Z}$$

Figure 14.10

Figure 14.11

Figure 14.12

203 × 203 UC60

Figure 14.13

Figure 14.15

EXERCISE 14

1 A short timber post carries a load of 40 kN eccentric from one axis only as shown in Fig. 14.10. Calculate the intensity of stress *a*) at face BC (f_{bc}), *b*) at face AD (f_{ad}).

2 Calculate the compression stress on the faces BC and AD of the short timber post, as shown in Fig. 14.11.

3 Repeat Question 2, but increasing the eccentricity to 60 mm.

4 Calculate the stress at the corner C of the short timber post shown in Fig. 14.12.

5 A short steel stanchion consists of a 203×203UC60 and carries loads as shown in Fig. 14.13. Calculate the stress at faces AD and BC.

6 A short steel post consisting of a 152×152UC30 carries three loads, as shown in Fig. 14.14. Calculate the stress at each of the corners *A*, *B*, *C* and *D*.

7 The short steel post, as shown in Fig. 14.15, is a 152×152UC37. There is a central point load of 120 kN and an eccentric load of *W* kN as shown. The maximum stress on the face BC is 80 N/mm². What is the value of the load *W*?

8 The vertical resultant load on the earth from a metre run of retaining wall is 40 kN applied as shown in Fig. 14.16. Calculate the intensity of vertical stress under the wall at A and B.

9 A triangular mass wall, as shown in section in Fig. 14.17, weighs 20 kN/m³ and rests on a flat base AB. What is the intensity of vertical bearing stress at A and B per length of wall?

Figure 14.14

Figure 14.16

Figure 14.17

Figure 14.18

Figure 14.20

Figure 14.19

10 The concrete beam shown in Fig. 14.18 is pre-stressed by the application of two loads of 300 kN applied at points 40 mm below the neutral axis of the section. *a*) What are the stresses at the upper and lower faces of the beam? *b*) At what distance below the neutral axis would the loads have to be applied for there to be no stress at all at the top face of the beam?

Note: Neglect self-weight of beam.

11 A double angle rafter member in a steel truss is subjected to an axial compression of 60 kN and also to bending moment from a point load of 10 kN, as shown in Fig. 14.19. The member is composed of two angles, as shown, for which the properties (given for the double angle section) are

$$A = 3000 \text{ mm}^2 \quad I_{xx} = 1.75 \times 10^6 \text{ mm}^4$$

Calculate the maximum compressive stress in the member.

12 A timber beam subjected to horizontal thrusts at each end is shown in Fig. 14.20, is loaded with a central concentrated load of 5 kN. The beam itself weighs 150 N. What is the maximum *a*) compressive stress? *b*) tensile stress?

13 A plain concrete beam, 240 mm × 600 mm, spans 6 m and weighs 3.2 kN/m. A horizontal compressive force of 900 kN is applied to the beam at 180 mm above its underside.

Calculate: *a*) the stresses in the top and bottom fibres of the beam; *b*) the additional udl the beam could carry without developing tension in its bottom fibres.

15 Gravity Retaining Walls

The previous chapter has been concerned mainly with the addition of direct and bending stress when these two types of stress occur within a material, e.g. the variation of stress across the face of a column section which arises due to eccentricity of loading.

The principles involved in those cases are used in much the same way on occasions where, for example, a wall, resting on soil or on a concrete footing and acted upon by horizontal forces, is transmitting to the soil or footing stresses which consist of

1 Direct stress from the wall's weight.

2 Stress due to the overturning moment.

Before these resultant stresses are considered, it will be necessary to study the effect of the combined action of the vertical and horizontal (or inclined) forces on the overall behaviour of the wall.

As a result of that action the wall may fail in three ways:

  i) sliding

 ii) overturning

iii) overstressing.

Since, however, the main purpose of a retaining wall is to provide resistance to the horizontal (or inclined) forces caused by the retained material, the nature of the pressures these forces exert on the wall will be investigated first.

HORIZONTAL FORCES: 1) WIND PRESSURE

This is the simplest case, because wind pressure is assumed to be uniform. Therefore, the total resultant force acts at the centre of the area over which the pressure is applied and this force, $P = p \times$ area newtons, or since, in the case of a wall,

Figure 15.1

(a)

(b)

Figure 15.2

Figure 15.3

1 m length of wall is generally considered,

$$P = p \times 1 \times H = p \times H \text{ newtons}$$

where p = unit wind pressure in N/m^2

H = the height of the part of the wall subject to that wind pressure (see Fig. 15.1) in metres.

The determination of the unit wind pressure p is detailed in CP3:Chapter V:Part 2:Wind loads. It is based on several factors, including not only the basic speed of the wind and the type of topography in the locality of the wall, but also a statistical factor which takes into account the probability of the basic wind speed being exceeded within the projected life span of the wall.

HORIZONTAL FORCES: 2) LIQUID PRESSURE

Consider the vertical surface AB, shown in Fig. 15.2(a), to be the face of a wall which is retaining a liquid. It can be shown that a cubic metre of liquid, situated at a depth h metres below the surface, exerts a pressure of $w \times h$ kN outwards on all its six side surfaces. w in this case is the equivalent density or the unit weight of the liquid in kN/m^3.

Thus the intensity of outward pressure varies directly with the depth and will have a maximum value of $w \times H$ kN/m^2 at H m, the maximum depth as indicated in Fig. 15.2(b).

At the surface of the liquid (where $h = 0$), the pressure will be zero. So, as the maximum is wH kN/m^2, the average pressure between A and B is $\frac{1}{2}wH$ kN/m^2.

In dealing with retaining walls generally, it is convenient, as was said earlier, to consider the forces acting on one metre length of wall, that is, an area of wall H m high and 1 m measured perpendicular to the plane of the diagrams (Fig. 15.2(a) and (b)). Thus, as the "wetted area" concerned is H m^2, the total force caused by water pressure on a one metre strip of wall is

"Wetted area" × Average rate of pressure = $H \times \frac{1}{2}wH$

$$= \frac{1}{2}wH^2 \text{ kN}$$

This total resultant force on the wall's vertical surface from the liquid is (as will be seen from Fig. 15.2(b)) the resultant of a large number of forces, which range from zero at the top to wH at the base. The resultant will therefore act at a point $\frac{1}{3}$rd of H from the base, as shown in Fig. 15.3.

Note: If the liquid does not reach the top of the wall, as, for example, in Fig. 15.4, then the resultant force is calculated with H as the depth of the liquid and not as the height

Figure 15.4

Figure 15.5

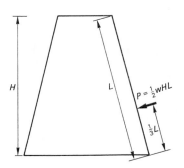

Figure 15.6

of the wall. The force is again $\frac{1}{2}wH^2$ and it acts at a point $\frac{1}{3}H$ (one-third the depth of the liquid) from the wall's base.

Example 15.1

A masonry dam retains water on its vertical face. The wall is, as shown in Fig. 15.5, 3.7 m, but the water level reaches only 0.7 m from the top of the wall. What is the resultant water pressure per metre run of wall?

Solution
The equivalent density w of water $= 10 \text{ kN/m}^3$
Therefore

$$P = \tfrac{1}{2}wH^2 = \tfrac{1}{2} \times 10 \times 3 \times 3 = 45 \text{ kN acting at 1 m above base}$$

In cases where the wall in contact with the water is not vertical, as is the case shown in Fig. 15.6, the wetted area will be larger than in the case of a vertical back, and the resultant pressure will thus be increased to $\frac{1}{2}wHL$ (i.e. the wetted area will be $L \text{ m}^2$ instead of $H \text{ m}^2$ considering one metre run of wall).

HORIZONTAL FORCES: 3) SOIL PRESSURE

It is obvious that pressures on walls from retained soil or other granular materials cannot be determined with quite the same accuracy as with water. Soils vary in weight and character; they behave quite differently under varying conditions of moisture, etc., and, in general, the resultant pressures on vertical and non-vertical surfaces from soils are obtained from various soil pressure theories. Numerous theories exist for the calculation of soil pressures, and these theories vary in the assumptions which they make and the estimated pressures which they determine. A great deal of research is still being directed upon this subject, but most of this is beyond the scope of a volume of this type, and therefore only the well tried Rankine's theory will be dealt with in detail.

Rankine's Theory of Soil Pressure

It has been seen that a cubic metre of liquid at a depth h below the surface presses outwards horizontally by an amount $wh \text{ kN/m}^2$ (w being the equivalent density of liquid). In the case of soil weighing $w \text{ kN/m}^3$, the outward pressure at

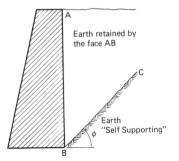

A

Earth retained by
the face AB

C

Earth
"Self Supporting"

ϕ

B

Figure 15.7

$P = \frac{1}{2}(wH^2)\frac{1-\sin\phi}{1+\sin\phi}$ kN

$\frac{1}{3}H$

Figure 15.8

a depth of h m below the surface will be less than wh kN/m^2, since some of the soil is "self-supporting."

Consider, for example, the soil retained by the vertical face AB in Fig. 15.7. If the retaining face AB was removed, then some of the soil would probably collapse at once, and in the course of time the soil would assume a line BC, as shown. The angle ϕ made between the horizontal and the line BC varies with different types of soil, and is called the **angle of repose** or *the angle of internal friction* of the soil. It can be said, therefore, that only part of the soil was in fact being retained by the wall and was exerting pressure on the wall. Thus, it follows that the amount of pressure on the wall from the soil depends upon the angle of repose for the type of soil concerned, and Rankine's theory states in general terms that the outward pressure per square metre at a depth of h m due to a level fill of soil is

$$wh\left[\frac{1-\sin\phi}{1+\sin\phi}\right] \text{kN/m}^2$$

as compared with (wh) kN/m^2 in the case of liquids. Thus, by similar reasoning as used in the case of liquid pressure, the maximum pressure at the bottom of the wall is given by

$$\text{Maximum pressure} = wH\left[\frac{1-\sin\phi}{1+\sin\phi}\right] \text{kN/m}^2$$

$$\text{Average pressure} = \tfrac{1}{2}wH\left[\frac{1-\sin\phi}{1+\sin\phi}\right] \text{kN/m}^2$$

The soil acts at this average rate on an area of H m^2 of wall, so the total resultant force per metre run of wall is

$$P = \tfrac{1}{2}wH^2\left[\frac{1-\sin\phi}{1+\sin\phi}\right] \text{kN}$$

and this acts, as shown in Fig. 15.8, at $\frac{1}{3}H$ above the base of the wall.

Example 15.2

Soil weighing 15 kN/m^3 and having an angle of repose ϕ of 30°, exerts pressure on a 4.5 m high vertical face. What is the resultant horizontal force per metre run of wall?

Solution $\sin\phi = \sin 30° = 0.5$ and

$$P = \tfrac{1}{2}wH^2\left(\frac{1-\sin\phi}{1+\sin\phi}\right)$$

$$= \tfrac{1}{2}\times 15\times 4.5\times 4.5\times\frac{1-0.5}{1+0.5}$$

$$= 151.875\times\tfrac{1}{3} = 50.625 \text{ kN}$$

Figure 15.9

MODES OF FAILURE: 1) SLIDING

It was stated at the beginning of the chapter that a retaining wall may fail by sliding. The possibility that the wall may slide along its base exists unless the weight of the wall is sufficient to prevent such movement.

The resistance to sliding depends upon the interaction of the weight of the wall and the friction between the material of the wall and the soil directly in contact with the base of the wall.

The basis of this interaction may be explained as follows. Consider a wooden block of weight W resting on a steel surface, as in Fig. 15.9(a). The steel surface presses upwards in reaction to the downward weight of the block, as shown.

If, however, a small load P (not sufficient to move the block) is also placed on the block, as in Fig. 15.9(b), then the resultant of the forces W and P will be R_1, and the resultant upward pressure R_2 from the steel surface will also be inclined as shown. This resultant upward pressure will be inclined at an angle ϕ, as indicated.

If the force P is now gradually increased, the angle ϕ will also increase, until at a certain load (which depends on the nature of the two surfaces in contact and on the weight W) the block will move horizontally.

The angle ϕ which the resultant upward thrust makes with the vertical at the stage where the block starts to slide is known as the *angle of friction* between the two surfaces.

The tangent of ϕ is P_M/W (see Fig. 15.9(c)), so that

$$\tan \phi = \frac{\text{Least force which will cause sliding}}{\text{Weight of block}} \qquad (1)$$

and $\tan \phi$ is known as the *coefficient of friction* for the two materials and is denoted by the letter μ (mu).

For most materials this coefficient of friction will vary between 0.4 to 0.7, and it will be appreciated from equation (1) that

P_M = the least force that will cause sliding

= $W \times$ coefficient of friction μ

In the case of retaining walls, P_M, the force which would cause sliding, can be calculated as $W \times$ coefficient of friction, and the horizontal force P of the retained material should not exceed approximately half of the force P_M.

Example 15.3

The masonry dam, shown in Fig. 15.10, retains water to the full depth, as shown. The coefficient of friction between the base of the wall and the earth underneath is 0.7. Check if the

Figure 15.10

wall is safe against sliding.

Solution

P = actual horizontal pressure on side of wall

$$= \tfrac{1}{2}wH^2 = \tfrac{1}{2} \times 10 \times 4^2 = 80 \text{ kN}$$

P_M = horizontal force which would just cause sliding

$$= 0.7 \times W = 0.7 \times \tfrac{1}{2}(1+3) \times 4 \times 18$$

$$= 0.7(144) = 100.8 \text{ kN}$$

The actual pressure (80 kN) exceeds half the value of P_M, and the factor of safety against sliding is

$$\frac{\mu W}{P} = \frac{100.8}{80} = 1.26 < 2$$

which is undesirable.

Figure 15.11

MODES OF FAILURE: 2) OVERTURNING

A retaining wall may have quite a satisfactory resistance to sliding, but the positive action of the horizontal forces may tend to overturn it about its toe (Fig. 15.11).

Equilibrium will be upset when the wall just rises off the ground at the heel B, the turning point being the toe, A. (It is assumed that sliding of the wall will not occur.) When the wall is just lifting off the ground, the overturning moment due to the force P is just balanced by the restoring or balancing moment due to the weight of the wall, the wall being balanced on the edge A. Since the inclination of AB to the horizontal is very slight, distances can be taken assuming the wall is vertical.

Taking moments about A,

$$P \times \text{distance AD} = W \times \text{distance AC}$$

or

Overturning moment = Restoring (balancing) moment

In practice it is usual to apply a factor of safety of 2, where

$$\frac{\text{Factor of safety}}{\text{against overturning}} = \frac{\text{Restoring (balancing) moment}}{\text{Overturning moment}}$$

and, therefore,

Overturning moment = $\tfrac{1}{2} \times$ restoring (balancing) moment

1 m run of wall

0.4 m

2.7 m

Area subjected to wind pressure = 2.7 m²

(a)

1.35 kN

19.44 kN

0.2 m

$\frac{1}{2}2.7 = 1.35$ m

(b)

Figure 15.12

Example 15.4

A long boundary wall 2.7 m high and 0.4 m thick is constructed of brickwork weighing 18 kN/m³. If the maximum wind pressure uniformly distributed over the whole height of the wall is 500 N/m², calculate the factor of safety against overturning neglecting any small adhesive strength between the brickwork and its base.

Solution

$$\text{Weight of 1 m run of wall} = 2.7 \times 1.0 \times 0.4 \times 18$$
$$= 19.44 \text{ kN}$$
$$\text{Wind pressure on 1 m run of wall} = 2.7 \times 1.0 \times 0.5$$
$$= 1.35 \text{ kN}$$

and this can be taken as acting at the centre of height of the wall for purposes of taking moments about O (Fig. 15.12(*b*)).

$$\text{Restoring moment} = 19.44 \times \tfrac{1}{2}0.4$$
$$= 3.888 \text{ kN m}$$
$$\text{Overturning moment} = 1.35 \times \tfrac{1}{2}2.7$$
$$= 1.823 \text{ kN m}$$

Therefore

$$\text{FS against overturning} = 3.888/1.823 = 2.13$$

It will be found that a satisfactory factor of safety against overturning is achieved if the resultant of the horizontal and vertical forces crosses the base of the wall within its "middle third", i.e. when no tensile stresses are allowed to develop in the wall.

This comes within the scope of the third mode of failure of retaining walls and is the topic of the following section of this chapter.

x

Resultant (P + W)

P

W

S

y

P

Figure 15.13

MODES OF FAILURE: 3) OVERSTRESSING

When the weight of a wall per metre *W*, and the resultant pressure from the soil or liquid *P*, have been calculated, these two forces may be compounded to a resultant, as in Fig. 15.13. It will be shown that the position along the base at which this resultant cuts (i.e. at S) has an important bearing on the stability of the wall and on the pressures exerted by the wall upon the earth beneath.

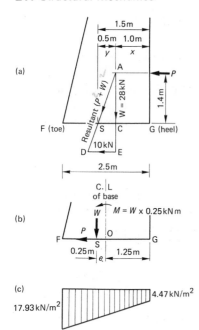

Figure 15.14

Consider now the wall shown in Fig. 15.14(*a*). The values in kN of *W* and *P* are as shown, and *W* acts through the centroid of the wall section.

From the similar triangles ASC and ADE

$$\frac{y}{1.4} = \frac{10}{28} \quad \text{so} \quad y = \frac{10 \times 1.4}{28} = 0.5 \text{ m}$$

Thus the resultant force cuts the base at $1.0 + 0.5 = 1.5$ m from the point G, the heel of the wall.

It can be seen that the resultant force cuts the base at point S, as shown in Figs. 15.14(*a*) and (*b*).

In considering the effect of this resultant on the soil or concrete under the base FG, it is normally convenient to resolve this force into the vertical and horizontal components *W* and *P* from which this resultant was compounded. (Note that the vertical component of the resultant is equal to the weight *W* of the wall only when the soil pressure is horizontal.)

These two component forces *W* and *P* are shown again in Fig. 15.14(*b*). Component force *P* tends to *slide* the wall along the plane FG, so need not be considered when calculating the pressure on the soil beneath FG. Component force *W*, however, does exert pressure on the soil or concrete under FG, and as the point S is to the left of the centre line of the base the soil under the left-hand part of the base will have greater pressure exerted upon it than the soil under the right-hand half. Here

$$\text{Direct stress} = \frac{W}{A} = \frac{\text{weight of wall}}{\text{area of base}}$$

$$= \frac{28}{2.5 \times 1.0} = 11.2 \text{ kN/m}^2$$

The moment due to the eccentricity of the resultant is

$W \times$ distance of S from the centre of the base

$$W \times 0.25 = 28 \times 0.25 = 7 \text{ kN m}$$

and the section modulus about an axis through the centre line of the base is

$$Z = \tfrac{1}{6}bd^2 = \tfrac{1}{6} \times 1 \times (2.5)^2 = 1.04 \text{ m}^3$$

$$\text{Bending stress} = \frac{M}{Z} = \frac{7}{1.04} = 6.73 \text{ kN/m}^2$$

Thus, the pressure under wall at F is

$$\frac{W}{A} + \frac{M}{Z} = 11.20 + 6.73 = 17.93 \text{ kN/m}^2$$

and the pressure under wall at G is

$$\frac{W}{A} - \frac{M}{Z} = 11.20 - 6.73 = 4.47 \text{ kN/m}^2$$

as shown in the stress or pressure distribution diagram of Fig. 15.14(c).

Example 15.5

A masonry wall is as shown in Fig. 15.15 and weighs 20 kN/m³. It retains on its vertical face water weighing 10 kN/m³. The water reaches the top of the wall. Calculate the pressures under the wall at the heel and the toe.

Figure 15.15

Solution

$$W = \tfrac{1}{2}(1.0 + 3.0) \times 4.5 \times 20 = 180.0 \text{ kN}$$
$$P = \tfrac{1}{2}wH^2 = \tfrac{1}{2} \times 10 \times (4.5)^2 = 101.25 \text{ kN}$$

To determine the centroid of the wall,

$$x = \frac{1.0 \times 4.5 \times 0.5 + \tfrac{1}{2} \times 2.0 \times 4.5 \times 1.7}{1.0 \times 4.5 + \tfrac{1}{2} \times 2.0 \times 4.5}$$

$$= \frac{2.25 + 7.50}{4.50 + 4.50} = 1.08 \text{ m}$$

By similar triangles

$$\frac{y}{1.5} = \frac{101.25}{180.0}$$

$$y = \frac{101.25 \times 1.5}{180.0} = 0.84 \text{ m}$$

Thus, the resultant force cuts at $1.08 + 0.84 = 1.92$ m from B at S or 0.42 m to the centre line of the base.

Hence, the pressure at A is

$$\frac{W}{A} + \frac{M}{Z} = \frac{180.0}{3.0 \times 1.0} + \frac{180.0 \times 0.42 \times 6}{1.0 \times 3.0 \times 3.0}$$

$$= 60.0 + 50.4 = +110.4 \text{ kN/m}^2$$

and the pressure at B is

$$\frac{W}{A} - \frac{M}{Z} = 60.0 - 50.4 = +9.6 \text{ kN/m}^2$$

The purpose of calculating distance y was to find the eccentricity of W, i.e. distance e, so that $M = We$ could be determined. Distance e, however, can be obtained directly by considering the equilibrium about the centroid of the base of

Figure 15.16

Figure 15.17

all the forces acting on the wall. This is illustrated in Fig. 15.16.

Let R_H and R_V be, respectively, the horizontal and vertical components of the soil's reaction acting at the intersection of the resultant of P and W with the base of the wall.

Then, for equilibrium, $R_H = P$, and $R_V = W$, both of which have been calculated previously.

Hence, taking moments about O, the centre of area (or centroid) of the base:

$$W \times a - P \times \tfrac{1}{3}H + W \times e = 0$$

$$e = \frac{P \times \tfrac{1}{3}H - W \times a}{W}$$

$$= \frac{101.25 \times 1.5 - 180.0 \times (1.50 - 1.08)}{180.0}.$$

$$= 0.42 \text{ m as before}$$

Example 15.6

The trapezoidal retaining wall shown in Fig. 15.17 weighs 22 kN/m^3 and retains on its vertical face soil with an equivalent density of 16 kN/m^3 and an angle of repose (or internal friction) ϕ of $30°$. The retained soil carries a superimposed vertical load of 9.6 kN/m^2. Determine the pressure under the base AB.

Solution The imposed load is converted to an equivalent additional height of soil as shown in Fig. 15.18 (*a*), where

$$h_1 = \frac{\text{Intensity of imposed load kN/m}^2}{\text{Equivalent density of soil kN/m}^3}$$

$$= \frac{9.6}{16.0} = 0.6 \text{ m}$$

Therefore, the pressure at the top of the wall, level T, is

$$w \times h_1 \times \frac{1 - \sin \phi}{1 + \sin \phi} = 16 \times 0.6 \times \tfrac{1}{3} = 3.2 \text{ kN/m}^2$$

and that on the underside of the wall, at level U, is

$$w \times (h_1 + H) \times \frac{1 - \sin \phi}{1 + \sin \phi} = 16 \times 4.6 \times \tfrac{1}{3} = 24.5 \text{ kN/m}^2$$

The force P, acting on the vertical face of the wall, is found from the shaded trapezoid of pressure and it acts at the centroid of the trapezoid, i.e.

$$P = \tfrac{1}{2}(3.2 + 24.5) \times 4 = 55.4 \text{ kN}$$

(a)

(b)

(c)

Figure 15.18

Figure 15.19

The distance d between the base of the wall and the line of action of force P is

$$d = \tfrac{1}{54}[3.2 \times 4 \times 2 + (24.5 - 3.2) \times \tfrac{4}{2} \times \tfrac{4}{3}]$$

$$= \frac{25.6 + 56.8}{55.4} = 1.49 \text{ m}$$

Now, the vertical force W due to the weight of the wall is

$$W = \tfrac{1}{2}(2.9 + 0.5) \times 4 \times 22 = 150 \text{ kN}$$

To find distance a between the line of action of force W and the centre line of the base, take moments about that centre line, then

$$a = \frac{0.5 \times 4 \times (1.45 - 0.25) + 2.4 \times \tfrac{1}{2}4 \times (1.45 - \tfrac{1}{3}2.4 - 0.5)}{0.5 \times 4 + 2.4 \times 2}$$

$$= \frac{2.4 + 0.72}{2.0 + 4.8} = \frac{3.12}{6.80}$$

$$= 0.46 \text{ mm}$$

Finally, to determine the eccentricity e, consider the moments of *all* the forces (i.e. the applied loads, P and W, and the reaction of the soil, R) about the centre line of the base, as demonstrated above. Therefore,

$$e = \frac{150 \times 0.46 - 55.4 \times 1.49}{150} = \frac{-13.5}{150}$$

$$= -0.09 \text{ m (i.e. to the left of centre line, Fig. 15.18}(b))$$

Thus, pressure at A is

$$\frac{W}{A} + \frac{We}{Z} = \frac{150}{1.0 \times 2.9} + \frac{150 \times 0.09 \times 6}{1.0 \times 2.9 \times 2.9}$$

$$= 51.72 + 9.63 = 61.35 \text{ kN/m}^2$$

and pressure at B is

$$51.72 - 9.63 = 42.09 \text{ kN/m}^2$$

The pressure distribution diagram is shown in Fig. 15.18(c).

The previous examples have been chosen so that the resultant force has cut the base within the middle third. Hence, although the pressures at B and A (the heel and toe of the base) have been of different amounts, the stresses have been compressive at both points. Consider, however, the case shown in Fig. 15.19. The resultant here cuts outside the middle third and thus there will be a tendency to tension at B, i.e. M/Z will be greater than W/A.

If the joint at B is capable of resisting tensile stress then the stress at A will be

Figure 15.20

(a)

(b)

(c)

Figure 15.21

Figure 15.22

$$\frac{W}{A}+\frac{M}{Z}=\frac{42}{1.0\times3.0}+\frac{42\times(1.5-0.75)\times6}{1.0\times3.0\times3.0}$$
$$=14.0+21.0=35 \text{ kN/m}^2$$

and the stress at B will be

$$\frac{W}{A}-\frac{M}{Z}=14.0-21.0$$
$$=-7.0 \text{ kN/m}^2 \text{ or } 7.0 \text{ kN/m}^2 \text{ tension}$$

The diagram of pressure distribution under the base will be as in Fig. 15.20.

If the joint at B is not capable of resisting tension, then there will be a tendency for the base to lift where tension would occur. In this case, the point at which the resultant cuts the base should be considered the middle third point, as in Fig. 15.21(b), so that the effective width of the base becomes three times the distance AS, i.e.

$$3\times0.75=2.25 \text{ m}$$

The stress at A (the toe) then becomes

$$\frac{W}{A}+\frac{M}{Z}=\frac{42}{1.0\times2.25}+\frac{42\times0.375\times6}{1.0\times2.25\times2.25}$$
$$=18.67+18.67=37.33 \text{ kN/m}^2$$

and the pressure distribution diagram is as shown in Fig. 15.21(c).

Example 15.7

For the retaining wall shown in Fig. 15.22 determine:
a) the position of the intersection point of the resultant of forces P and W with the base AB when soil is level with top of wall,
b) the reduced height of the soil which will cause the point of intersection to coincide with the edge of the middle third of the base.

Assume: w for wall $= 20$ kN/m^3
  w for soil $= 15$ kN/m^3
  ϕ for soil $= 30°$.

Solution a)

$$W=\tfrac{1}{2}(1.0+2.5)\times6\times20=210 \text{ kN}$$
$$P=\tfrac{1}{2}(15\times6^2)\times\tfrac{1}{3}=90 \text{ kN}$$

To determine the values of a and e take moments about centre line of base.

$$a = \frac{1.0 \times 6.0 \times 0.75 - 1.5 \times \frac{1}{2}6.0 \times 0.25}{\frac{1}{2}(1.0 + 2.5) \times 6.0}$$

$$= \frac{4.5 - 1.125}{10.5} = 0.32 \, \text{m to right of centre}$$

$$e = \frac{90 \times \frac{1}{3}6.0 - 210 \times 0.32}{210} = \frac{180 - 67.5}{210}$$

$$= 0.54 \, \text{m} > 2.5/6 \, \text{i.e. tension will develop}$$

Solution b)

$$\frac{1}{2}(15 \times H^2) \times \frac{1}{3}H \times \frac{1}{3} - 210 \times (0.32 + \frac{1}{6}2.5) = 0$$

$$H^3 = \frac{(67.5 + 87.5) \times 18}{15} = 186$$

$$H = 5.71 \, \text{m}$$

The above example may be used quite conveniently to verify the final statement in the previous section on Overturning. Check the stability of the above wall in both cases, *a*) and *b*).

Moments are taken about the toe of the wall at A.

a) $\dfrac{\text{Restoring moment}}{\text{Overturning moment}} = \dfrac{210 \times (1.25 + 0.32)}{180}$

$$= 1.83 < 2 \text{ unsatisfactory}$$

b) $\dfrac{210 \times 1.57 \times 18}{15 \times (5.71)^3} = 2.13 \text{ satisfactory}$

Finally, it should be noted that the main point of calculating the pressures under the base of the wall is to ensure that the soil on which the wall is founded is not overstressed. It is, therefore, essential to keep the actual pressures within the limits of the safe bearing capacities of the soil.

The safe bearing capacities vary over a very wide range of values from, say, $50 \, \text{kN/m}^2$ for made-up ground to $650 \, \text{kN/m}^2$ for compact, well graded sands or stiff boulder clays. The values for rocks are, of course, higher.

SUMMARY

The three **modes of failure** of a retaining wall are

1) Sliding
2) Overturning
3) Overstressing

The **horizontal force,** P, acting on 1 m length of the vertical face of a retaining wall is

1) Wind $\quad p \times H$
2) Liquid $\quad \frac{1}{2}wH^2$
3) Soil (or granular material)

$$\frac{1}{2}wH^2 \times \frac{1-\sin\phi}{1+\sin\phi} \quad \text{Rankine's Formula}$$

where p = unit wind pressure in N/m^2
$\quad\quad w$ = equivalent density (or unit weight) of retained material in kN/m^3
$\quad\quad \phi$ = angle of repose (or internal friction) of soil (or granular material)
$\quad\quad H$ = height (or depth) or retained material (wind included) in metres.

Resistance to **sliding**

$$\frac{W \times \mu}{P_M} = \text{Factor of Safety (usually 2)}$$

where W = weight of wall in kN
$\quad\quad \mu$ = coefficient of friction between wall and supporting soil (0.4–0.7)
$\quad\quad P_M$ = the least force that will cause sliding, in kN.

Resistance to **overturning**

$$\frac{\text{Restoring (balancing) moment}}{\text{Overturning moment}} = \text{Factor of Safety (2)}$$

[Moments are taken about an outer edge (toe or heel) of the base of the wall.]

Resistance to **overstressing**
The maximum stress resulting from the combination of direct and bending stresses must be kept within limits of the safe bearing capacity of the soil supporting the wall.

For "no tension", the resultant of the applied loads P and W must cross the base within its middle third or

$$e \not> \tfrac{1}{6} \times \text{width of base}$$

Figure 15.23

EXERCISE 15

1 Calculate the resultant horizontal force P acting on the wall shown in Fig. 15.23 for
a) a liquid of equivalent density $w = 10 \text{ kN/m}^3$,
b) a soil of equivalent density $w = 16 \text{ kN/m}^3$ and angle of repose $\phi = 30°$, given that $H =$ i) 2 m, ii) 3 m, iii) 7 m.

2 The walls shown in Fig. 15.24(*a*) and (*b*) weigh 22 kN/m³. Calculate the value of P in kN per m run of wall, if there is to be a factor of safety of 2 against sliding. $\mu = 0.6$.

3 The horizontal forces on the walls shown in Fig. 15.25(*a*), (*b*) and (*c*) represent the forces on a 1 m run of wall. All the walls weigh 22 kN/m³. Calculate the factor of safety against overturning of the walls. (*Note:* only the position of the vertical line in which the c.g. is situated need be calculated.)

4 The wall, shown in Fig. 15.26, has a resultant vertical force of 57 kN/m length, due to the weight of the wall W and the pressure P. Calculate pressure at the toe (point A).

5 A masonry retaining wall is trapezoidal, being 4 m high, 1.4 m wide at the top and 2.0 m at the base. The masonry weighs 20 kN/m³ and the wall retains on its vertical face earth weighing 15 kN/m³ and having an angle of repose of 40°. Calculate the maximum pressure under the wall.

Figure 15.24

Figure 15.25

Figure 15.27

Figure 15.26

6 A brick pier H m high is 450 mm × 300 mm in section and weighs 18 kN/m³. It has a uniform wind pressure of 720 N/m² on one side, as shown in Fig. 15.27 and rests on a concrete block to which it is not connected. How high is the wall if the resultant force cuts through the point A?

Figure 15.28

Figure 15.29

Figure 15.30

7 A concrete wall of trapezoidal section, 4.8 m high, has a top width of 1.0 m and a base width of 2.5 m with one face vertical and a uniform batter on the other face. There is water pressure on the vertical face, with top water level 1 m below the top of the wall. If the water weighs 10 kN/m³ and the masonry 24 kN/m³, calculate the maximum and minimum pressures on a horizontal plane 2 m above the base of the wall.

8 A brick wall 450 mm thick and 3.0 m high is built on a level solid concrete base slab. The wall weighs 20 kN/m³. Assuming no tension at the bed joint, calculate the intensity of the uniformly distributed horizontal wind force in N/m² over the full height of the wall that would just cause overturning of the wall.

9 A concrete retaining wall is as shown in Fig. 15.28. Using the Rankine formula, calculate the pressure under the wall at A and B. Assume the soil to weigh 16 kN/m³, and the natural angle of repose of the soil to be 30°. The concrete weighs 24 kN/m³.

10 Fig. 15.29 shows the section of a mass concrete retaining wall, weighing 24 kN/m³. The position and amount of the resultant horizontal force is as shown. Calculate the vertical pressures under the wall at A and B, assuming *a*) tension stresses are permitted, *b*) no tension stresses can develop.

11 A small mass concrete retaining wall, 2 m high, is trapezoidal in section, 0.3 m wide at the top and 1.2 m wide at the base. The wall weighs 24 kN/m³ and it retains on its vertical face soil weighing 14 kN/m³ and having an angle of repose of 30°. Calculate the pressure under the base of the wall at the heel and at the toe.

12 A wall 0.23 m thick, weighing 19 kN/m³, rests on a solid foundation. It has a uniform horizontal pressure of 0.7 kN/m² on one face. If the resultant force at the base just cuts at the edge of the wall (i.e. if overturning is just about to take place), how high is the wall?

13 How high will a wall, similar to that mentioned in Question 12, be if the resultant pressure at the base cuts at the middle third, i.e. if tension is about to occur?

14 An L-shaped retaining wall is shown in Fig. 15.30. The masonry weighs 22 kN/m³. Calculate the maximum and minimum pressures under the wall in kN/m².

Answers to Exercises

EXERCISE 2

1 (*a*) 51.8 N, 73.2 N; (*b*) 183.0 N, 224.2 N; (*c*) 577.4 N,
288.7 N   **2** 14.5°   **3** $L = 0.87$ kN, $R = 0.50$ kN, Reaction =
1.36 kN   **4** $L = 2.04$ kN, $R = 0.75$ kN, $W = 2.88$ kN, $M = 1.06$ kN
5 19.9 kN at 31° to the vertical   **6** 7.3 kN at 25° to the
vertical   **7** 9.6 kN at 81° to the vertical   **8** A = 32°, resultant =
17.2 kN   **9** 7.3 kN at 70° to the vertical   **10** 559 N at 23° to the
vertical   **11** 677 N at 17° to the vertical   **12** 774 N at 64°, 554 N at
22.5°, 793 at 37.5° to the vertical   **13** (*a*) 35 kN, 40 kN; (*b*) 8 kN,
8 kN; (*c*) 17 kN, 10 kN   **14** 9.2 kN, $x = 22.5°$, 8.5 kN, 3.5 kN
15 $L = 2.5$ kN, $L_v = 1.3$ kN, $L_h = 2.2$ kN; $R = 4.3$ kN, $R_v = 3.7$ kN,
$R_h = 2.2$ kN   **16** 5.5 kN, 6.7 kN, 4.8 kN, $R_L = 4.8$ kN, $R_R = 2.8$ kN
17 $T = S = 5$ kN, $A_v = B_v = 4.3$ kN, $A_h = B_h = 2.5$ kN
18 AB = 173 N, BC = 100 N, $A_v = 150$ N, $A_h = 87$ N   **19** 5.0 kN,
6.1 kN, 6.8 kN   **20** 82.5°, 6.1 kN   **21** 2.9 kN
22 $A = 9.0$ kN, $B = 7.3$ kN   **23** 0.29 kN
24 4.0 kN in inclined members, 2.3 kN in horizontal member and
in cable, vertical reactions = 2 kN   **25** BC = CD = 3.0 kN, BD = 3.5 kN,
AB = AD = 1.7 kN, $R_A = 3.0$ kN   **26** (*a*) $X = 64.6$ N down,
$Y = 82.0$ N up; (*b*) $X = 4.6$ N down, $Y = 22.0$ N up; (*c*) $X = 1.6$ kN up,
$Y = 3.2$ kN down; (*d*) $X = 0.4$ kN down, $Y = 1.2$ kN down;
(*e*) 38.6 kN at 43° to vertical; (*f*) 41.4 kN at 40° to the vertical
27 13.7 kN, 11.8 kN.

EXERCISE 3

1 $A = 5.30$ kN, $B = 3.95$ kN   **2** $R_A = 17.3$ kN, $R_B = 20.0$ kN, $AC = 20.0$ kN, $BC = 17.3$ kN, $AB = 10.0$ kN, $R_{Bv} = 10.0$ kN, $R_{Bh} = 17.3$ kN   **3** Top = 385 N, bottom = 631 N (385 N horizontal, 500 N vertical)   **4** Cable 1.9 kN, hinge 1.9 kN   **5** Cable 1.4 kN, hinge 1.4 kN   **6** $A = 0.50$ kN, $B = 0.87$ kN   **7** $A = 0.67$ kN, $B = 1.20$ kN (1.00 kN vertical, 0.67 kN horizontal)   **8** String 1000 N, hinge 900 N   **9** Rope 6.1 kN, hinge 6.0 kN   **10** 500 N, 500 N   **11** Cable 1.15 kN, hinge 1.53 kN   **12** Rope 14.1 kN, hinge 22.4 kN   **13** Rope 3.5 kN, hinge 7.9 kN   **14** 3.87 kN, 8.24 kN, 8.87 kN   **15** $A = 7.64$ kN, $B = 5.77$ kN, $AB = 10.0$ kN, $AC = 2.75$ kN, $BC = 5.77$ kN   **16** $x = 20°$ **17** 354 N, 250 N   **18** $A = 334$ N, $B = 213$ N   **19** $A = 315$ N, $B = 192$ N   **20** $A = 0.37$ kN, $B = 0.78$ kN   **21** 433 N, 578 N, 807 N   **22** 4.72 kN, 4.25 kN, 2.51 kN   **23** $A = 6.3$ kN, $B = 2.9$ kN   **24** (a) 62.5 N at 70° to the vertical, 100 mm from bottom r.h. corner; (b) 60.0 N at 65° to the vertical, 600 mm from bottom l.h. corner; (c) 75.0 N at 48° to the vertical, 1.1 m from bottom l.h. corner   **25** 3.8 kN at 72° to the vertical, 1.7 m from bottom   **26** 6.0 kN, 6.0 m from bottom   **27** 38.5 kN, 41.0 kN at 32.5° to the vertical   **28** 60.0 kN, 118.6 kN at 37.5° to the vertical   **29** $X = 11.4$ kN, $Y = 6.5$ kN at 6.6° to the vertical   **30** 50.0 kN, 22.4 kN at 63.4° to the vertical   **31** 55.0 kN, 65.0 kN   **32** 4.4 kN, 11.5 kN at 41° to the vertical   **33** $R = 5.7$ kN at 41° to the vertical, $L = 9.4$ kN at 23° to the vertical   **34** 8.2 kN, 16.8 kN at 80° to the vertical   **35** 11.6 kN at 45° to the vertical, 8.7 kN at 71° to the vertical   **36** 43.3 kN, 49.8 kN at 20.5° to the vertical   **37** $R = 44.2$ kN at 11.0° to the vertical, $L = 47.5$ kN at 10.5° to the vertical   **38** 13.2 kN, 15.4 kN at 27° to the vertical **39** $R = 40.5$ kN at 7° to the vertical, $L = 57.0$ kN at 5° to the vertical.

EXERCISE 4

1 110 N, 205 N   **2** 1.5 m   **3** 375 N, 225 N (upwards)   **4** 35 mm, 265 N   **5** 0.23 m   **6** 1.64 kN, 1.06 kN   **7** 4.23 kN, 1.53 kN (downwards)   **8** $W = 22.5$ N, $A = 425.0$ N, $B = 237.5$ N   **9** See Answers to Exercise 3   **10** 1.44 kN, 2.36 kN at 17.7° to the vertical   **11** 2.5 kN, 1.8 kN at 65.3° to the vertical   **12** 30.6 kN, 34.2 kN, 25.0 kN, 43.3 kN   **13** 6.0 kN, 5.34 kN downwards at 76.5° to the vertical   **14** 3.9 kN, 6.9 kN   **15** (a) 0.41 m from hinge, 70 N, 176 N at 20° to the vertical; (b) 2.41 m from hinge, 4.12 kN, 6.10 kN at 42.5° to the vertical; (c) 2.27 m from hinge, 3.0 kN, 5.78 kN at 21.5° to the vertical   **16** (a) 2.33 m from hinge, 3.5 kN, 5.7 kN at 38° to the vertical; (b) 1.8 m from the hinge, 30 kN, 39 kN at 40° to the vertical; (c) 3.17 m from the hinge, 15.8 kN, 21.8 kN at 46.5° to the vertical; (d) 3.0 m from the hinge, 10.0 kN, 22.4 kN at 27° to the vertical   **17** 15.0 kN, 9.92 kN at 41° to the vertical, 6.50 kN   **18** 300 mm, 45 000 N mm   **19** (a) 50 kN, 70 kN; (b) 110 kN, 130 kN; (c) 120 kN, 10 kN; (d) 55 kN, 245 kN; (e) 70 kN, 150 kN   **20** (a) 225 kN, 135 kN; (b) 199 kN, 249 kN; (c) 324 kN, 108 kN; (d) 172.8 kN, 115.2 kN; (e) 221.5 kN, 60.5 kN; (f) 242.8 kN, 107.2 kN; (g) 523.1 kN, 231.9 kN   **21** $A = 40$ kN, $B = 50$ kN, $C = 40$ kN, $D = 10$ kN, $E = 26.7$ kN, $F = 13.3$ kN   **22** 14.07 kN.

EXERCISE 5

All forces are in kN units.

1

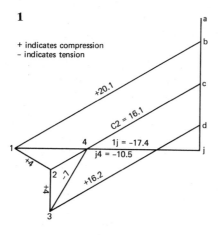

+ indicates compression
- indicates tension

+20.1
C2 = 16.1
1j = −17.4
j4 = −10.5
+16.2
+4
+4

2

+ Indicates compression
- Indicates tension

+21.3
+15.0
+7.0
−5.0
e1 = −15.0
e3 = −20.0

3

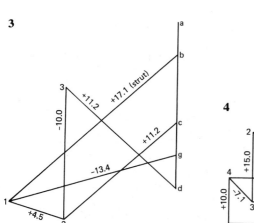

+11.2
+17.1 (strut)
+11.2
−10.0
−13.4
+4.5

4

+15.0
−21.2
+15.0
+20.0
+10.0
−7.1
−15.0
+20.0

5

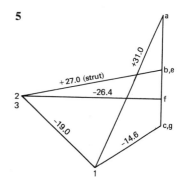

+27.0 (strut)
−26.4
+31.0
−19.0
−14.6

6

−38.0 (tie)
−29.0
+60.4
+60.8
−25.4
−35.0

7

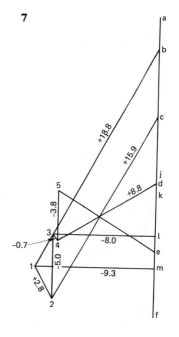

+18.8
+15.9
+8.8
−3.8
−0.7
−8.0
−5.0
−9.3
+2.8

8

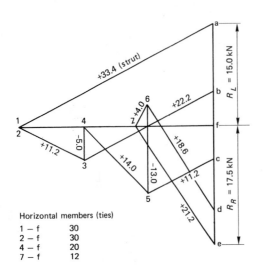

Horizontal members (ties)

1 – f	30
2 – f	30
4 – f	20
7 – f	12

10

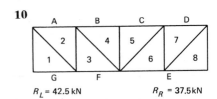

R_L = 42.5 kN R_R = 37.5 kN

9

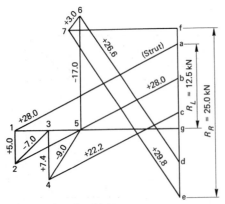

Horizontal members (ties)

1 – g	25.2
3 – g	20.0
5 – g	15.0
7 – f	17.0

11

12

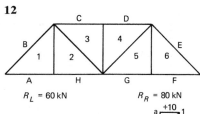

R_L = 60 kN R_R = 80 kN

14

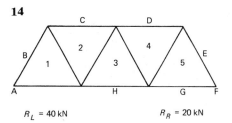

R_L = 40 kN R_R = 20 kN

13

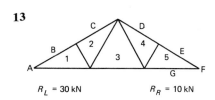

R_L = 30 kN R_R = 10 kN

15

16

Reaction at (y) = 32 kN
Reaction at (x) is given by ca
= 44 kN

17

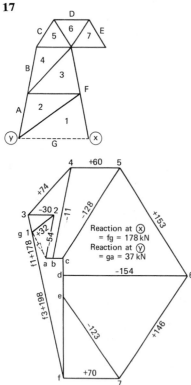

Reaction at (x)
= fg = 178 kN
Reaction at (y)
= ga = 37 kN

18

Reaction at (y) = ea = 104 kN

Reaction at (x) = de = 120 kN

19

Reaction at (x) = de = 67 kN
Reaction at (y) = ea = 60 kN

20

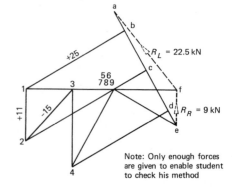

R_L = 22.5 kN

R_R = 9 kN

Note: Only enough forces
are given to enable student
to check his method

21

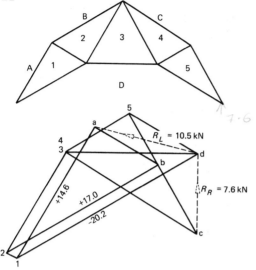

R_L = 10.5 kN

R_R = 7.6 kN

+14.6
+17.0
−20.2

23

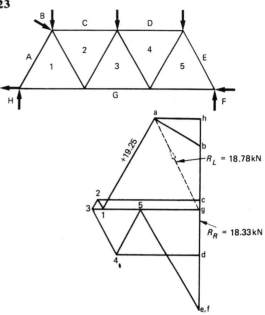

+19.25

R_L = 18.78 kN

R_R = 18.33 kN

22

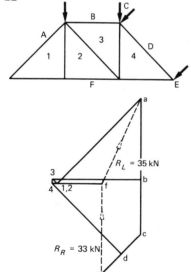

R_L = 35 kN

R_R = 33 kN

24

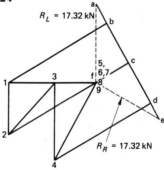

R_L = 17.32 kN

R_R = 17.32 kN

Notation as for Q20

25

R_L = 2.7 kN

R_R = 12.7 kN

Notation as for Q21

26

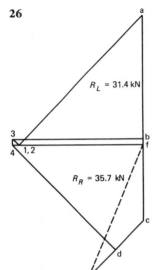

R_L = 31.4 kN

R_R = 35.7 kN

Notation as for Q22

27

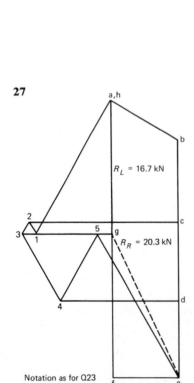

R_L = 16.7 kN

R_R = 20.3 kN

Notation as for Q23

28

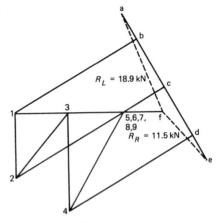

R_L = 18.9 kN

5,6,7, 8,9

R_R = 11.5 kN

Notation as for Q20

29

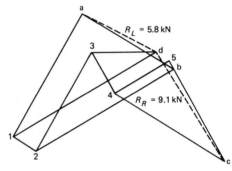

R_L = 5.8 kN

R_R = 9.1 kN

Notation as for Q21

30

R_L = 32.2 kN

R_R = 33.5 kN

Notation as for Q22

31

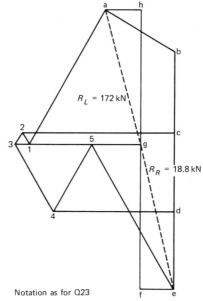

Notation as for Q23

32 AB = 2.39 kN ⎫
   AC = 0.98 kN ⎬ all struts
   AD = 2.25 kN ⎭
33 AB = 16.62 kN   strut
   AC = 9.82 kN   tie
   AD = 7.25 kN   tie
34 AB = AC = 0.7 kN   ties
   AD = AF = 2.43 kN   struts
   AE = 3.33 kN   tie
   BC = 0.5 kN   strut
   BF = DC = zero

EXERCISE 6

1 135 N/mm² **2** 69.75 kN **3** 8.9 mm **4** 122.2 N/mm² **5** 157.8 kN
6 18.5 mm **7** 80 kN, 26 mm **8** 55 mm,
99 kN **9** 9.3 mm **10** 460 mm **11** 5 **12** 442 mm
13 (a) 68 N/mm², (b) 516 mm, (c) 2.45 m **14** (a) 251 N/mm²,
(b) 404 N/mm², (c) 176 839 N/mm² **15** 1.5 mm **16** 141 N/mm²,
2 mm **17** 8727 N/mm², 1.6 mm **18** 146.25 kN, 1.6 mm **19** Copper
0.27 mm, brass 0.29 mm **20** 15.5 N/mm², 0.001,
15 500 N/mm² **21** 204.4 kN, 2 mm **22** 2×6.6 mm **23** Timber
5.0 N/mm², steel 131.7 N/mm², 1.9 mm **24** 70.2 kN, steel
109.2 N/mm², brass 41.6 N/mm².

EXERCISE 7

1 $L = 91$ kN, $R = 59$ kN, $M_{max} = 76$ kN m 1.03 m from L **2** $L = 145$ kN, $R = 135$ kN, $M_{max} = 151.9$ kN m 1.75 m from L **3** $L = 100$ kN, $R = 60$ kN, $M_{max} = 83.3$ kN m 1.67 m from L **4** $L = 24$ kN, $R = 36$ kN, $M_{max} = 57.6$ kN m 2.8 m from L **5** $L = 31.5$ kN, $R = 73.5$ kN, $M_{max} = +16.5$ kN m 1.05 m from L and -15.0 kN m at R
6 $L = 140.4$ kN, $R = 114.6$ kN, $M_{max} = 219.1$ kN m 3.18 m from L
7 $L = 99.6$ kN, $R = 104.4$ kN, $m_{max} = +14.3$ kN m 1.79 m from L and -35.0 kN m at R **8** $L = R = 60$ kN, $M_{max} = 96$ kN m at centre of span **9** $L = 112.1$ kN, $R = 27.9$ kN, $M_{max} = +7.38$ kN m 0.87 m from L and -30.0 kN m at L **10** $L = 120$ kN, $R = 200$ kN, $M_{max} = +120$ kN m between 1 m and 5 m from L and -120 kN m at R
11 $L = 97.5$ kN, $R = 187.5$ kN, $M_{max} = +35.1$ kN m 0.25 m from L and -120.0 kN m at R **12** $L = 75$ kN, $R = 175$ kN, $M_{max} = +5$ kN m 1 m from L and -140 kN m at R **13** $L = 169$ kN, $R = 201$ kN, $M_{max} = 282.95$ kN m 2.04 m from L **14** $L = 230$ kN, $R = 140$ kN, $M_{max} = 512.88$ kN m 4.5 m from L **15** $M_{max} = +28.9$ kN m and -80.0 kN m **16** 37.63 kN m **17** 2.83 m ($\sqrt{8}$ m) **18** (a) $L = 150$ kN, $R = 145$ kN; (b) $+158.75$ kN m 4.5 m from L, -70 kN at L **19** (a) $L = 142$ kN, $R = 148$ kN; (b) 181.4 kN m 2.4 m from L **20** $R = 10$ kN, $M_{max} = 30$ kN m at support **21** Max. SF $= 146.7$ kN at R, $M_{max} = 136.67$ kN m 2.5 m from L **22** (a) $L = 215.5$ kN, $R = 235.5$ kN; (b) $M_{max} = 708.25$ kNm 5.5 m from L **23** (a) 15.0 kN m; (b) 245.0 kN m; (c) 357.5 kN m **24** 400 kN m **25** (a) $R_L = 20$ kN, $R_R = 40$ kN; (b) $M_{max} = 46.20$ kN m; (c) $M_C = 31.11$ kN m
26 (a) $+20$ kN m; (b) -10 kN m; (c) $+5$ kN m **27** 120 kN **28** 6 m
29 60 kN **30** (a) $\frac{1}{2}Wl$; (b) $\frac{3}{8}Wl$.

EXERCISE 8

1 444 mm east and 444 mm south of A
2 (a) 22.3 mm; (b) 55.9 mm; (c) 82.1 mm; (d) 112.5 mm
3 (a) $\bar{x} = 8.3$ mm, $\bar{y} = 32.5$ mm; (b) $\bar{x} = 134.7$ mm, $\bar{y} = 62.9$ mm; (c) $\bar{x} = 76.3$ mm, $\bar{y} = 19.1$ mm; (d) $\bar{x} = 83.8$ mm, $\bar{y} = 148.5$ mm; (e) $\bar{x} = 55.0$ mm, $\bar{y} = 41.7$ mm
4 $I_{xx} = 381.24 \times 10^6$ mm^4, $I_{yy} = 47.50 \times 10^6$ mm^4
5 $I_{xx} = 247.30 \times 10^6$ mm^4, $I_{yy} = 126.25 \times 10^6$ mm^4
6 $a = 29.4$ mm, $I_{xx} = 4.20 \times 10^6$ mm^4, $I_{yy} = 9.54 \times 10^6$ mm^4
7 $a = 53.8$ mm, $I_{xx} = 11.6 \times 10^6$ mm^4, $I_{yy} = 2.15 \times 10^6$ mm^4
8 $a = 49.5$ mm, $b = 24.5$ mm, $I_{xx} = 6.56 \times 10^6$ mm^4, $I_{y.} = 2.37 \times 10^6$ mm^4
9 $a = 398$ mm, $I_{xx} = 998.79 \times 10^6$ mm^4, $I_{yy} = 76.00 \times 10^6$ mm^4
10 $a = 71$ mm
11 $I_{xx} = 12.13 \times 10^9$ mm^4, $I_{yy} = 3.19 \times 10^9$ mm^4
12 $x = 127.7$ mm, $I_{xx} = 199.34 \times 10^6$ mm^4
13 $x = 90.5$ mm, $I_{xx} = 117.51 \times 10^6$ mm^4
14 $x = 49.5$ mm, $I_{xx} = 6.56 \times 10^6$ mm^4
15 $I_{xx} = 37.83 \times 10^6$ mm^4, $I_{yy} = 9.02 \times 10^6$ mm^4
16 $I_{xx} = I_{yy} = 3.54 \times 10^6$ mm^4
17 $x = 37$ mm, $I_{xx} = 161.97 \times 10^6$ mm^4, $I_{yy} = 74.73 \times 10^6$ mm^4
18 $I_{xx} = 159.31 \times 10^6$ mm^4, $I_{yy} = 44.95 \times 10^6$ mm^4
19 24 mm **20** 160 mm **21** 24 mm
22 221 mm **23** $I_{xx} = 2.69 \times 10^9$ mm^4.

EXERCISE 9

1 25 kN **2** 258 mm **3** 409 kN **4** 171 kN **5** 148 N/mm² **6** 96 kN
7 6.8 kN **8** 8.8 kN **9** 8.88 kN/m² **10** (a) 4.86 kN/m²;
(b) 0.156×10^6 mm³ **11** (a) 137 mm; (b) $I_{xx} = 280.9 \times 10^6$ mm⁴,
$Z_{xxC} = 1.26 \times 10^6$ mm³, $Z_{xxT} = 2.05 \times 10^6$ mm³; (c) 25.3 kN/m
12 453 mm **13** (a) & (f) 254 × 148UB37; (b) & (c)
305 × 102UB33; (d) 533 × 210UB101; (e) 457 × 152UB74
14 (a) 62.5 N/mm²; (b) 21.4 N/mm² **15** 203 × 133UB25
16 Joists: 142 mm × 75 mm, 133 mm × 48 mm; A = 305 × 102UB28,
B = 254 × 102UB22 **17** l = 6.4 m, beam = 356 × 127UB39
18 22 m **19** 406 × 178UB74 **20** 305 × 102UB28
21 (a) 102 kN; (b) 95 mm **22** 10.5 kN.

EXERCISE 10

1 (a) (i) 120 N/mm², (ii) 8 N/mm²; (b) 81.6×10^6 N mm
2 99.5 kN **3** 74.8×10^6 N mm **4** (a) 44.9 kN; (b) 20 mm
5 592 mm **6** d_1 = 127 mm, A_{st} = 475 mm² **7** (a) 590 mm;
(b) 1520 mm² **8** d_1 = 289 mm, A_{st} = 776 mm² **9** d_1 = 380 mm,
A_{st} = 1246 mm², W = 12.5 kN **10** d_1 = 318 mm, A_{st} = 856 mm²
11 d_1 = 200 mm, A_{st} = 844 mm² **12** Beam A: d_1 = 323 mm,
A_{st} = 865 mm²; beam B: d_1 = 167 mm, A_{st} = 224 mm² **13** d_1 = 430 mm,
A_{st} = 1879 mm².

EXERCISE 11

1 18 mm **2** 14 mm **3** 16 mm **4** 108 kN **5** 6 mm **6** 1.5 mm
7 1.4 mm **8** 7.5 mm **9** 8.3 mm **10** 14.5 mm **11** (a) 8.7 kN;
(b) 3.3 kN **12** (a) 225 mm; (b) 309 mm **13** 12 mm **14** 7.5 mm
15 75 mm × 155 mm **16** 7.5 **17** 127 mm **18** 18 mm
19 200 mm × 600 mm **20** 600 mm, 198.7 kN **21** 10 mm.

EXERCISE 12

1 (a) 15.9 kN; (b) 30.7 kN **2** (a) 5.3 kN; (b) 11.4 kN;
(c) 139.3 kN **3** 200 mm **4** 143.3 kN **5** 420 mm **6** 62.4 kN,
120.6 kN **7** (a) 604.2 kN; (b) 702.2 kN; (c) 1014.6 kN **8** 11.6 m,
269.4 kN **9** 203 × 203UC71 **10** (a) 681 kN; (b)
782 kN **11** 565 kN, 120 mm **12** 54.0 kN, 31.2 kN,
14.8 kN **13** 1.3 MN **14** 1.6 MN **15** 422 mm square,
7117 mm² **16** (a) 370 mm square, 10 994 mm²; (b) 690 mm square,
3800 mm² **17** 582.3 kN, 1007.2 kN **18** 250 mm × 290 mm,
2900 mm² **19** 330 mm square, 4356 mm²
20 980 kN **21** 6597 mm² **22** (a) 200 mm square,
2000 mm²; (b) 4.0 m.

EXERCISE 13

1 145 kN (plate)   **2** 152 kN (rivets)   **3** 206.15 kN (plate)
4 164.3 kN (plate)   **5** (a) 124 kN (plate); (b) 148.8 kN (plate)
6 8.3 mm, 4 bolts each in plate and column   **7** 3.2 (i.e. 4) bolts,
134 kN (see Cl.42 BS 449)   **8** $x = 70$ mm, 5 bolts   **9** AA 245.5 kN,
BB 262.9 kN, CC 282.7 kN, rivets 254.5 kN   **10** $W = 163$ kN,
21 mm, 8 rivets   **11** 4 bolts, 75 mm (say)   **12** 200 kN.

EXERCISE 14

1 (a) $+6.88$ N/mm^2; (b) $+1.12$ N/mm^2 **2** $f_{bc} = 6.4$ N/mm^2, $f_{ad} = 0$
3 $f_{bc} = +7.04$ N/mm^2, $f_{ad} = -0.64$ N/mm^2 **4** 7.81 N/mm^2 **5** AD =
$+90$ N/mm^2, BC $= +47$ N/mm^2 **6** $A = +10$ N/mm^2, $B = -6$ N/mm^2,
$C = +158$ N/mm^2, $D = +174$ N/mm^2 **7** 63 kN   **8** $A = +8.67$ kN/m^2,
$B = +4.67$ kN/m^2 **9** $A = 52.1$ kN/m^2, $B = 0$ **10** (a) $+12$ N/mm^2 (bottom), $+1.33$ N/mm^2 (top); (b) 50 mm   **11** 72 N/mm^2
12 (a) $+8.0$ N/mm^2; (b) -5.6 N/mm^2 **13** (a) -0.25 N/mm^2 (top),
$+12.75$ N/mm^2 (bottom); (b) $W = 19.2$ kN.

EXERCISE 15

1 (a) (i) 20 kN/m^2, (ii) 45 kN/m^2, (iii) 245 kN/m^2; (b) (i) 10.7 kN/m^2,
(ii) 24 kN/m^2, (iii) 131 kN/m^2 **2** (a) 69.3 kN; (b) 29.7 kN   **3** (a) 8.3;
(b) 9.3; (c) 8.2   **4** 35.42 kN/m^2 **5** 80 kN/m^2 **6** 5 m   **7** 66.1 kN/m^2,
36.9 kN/m^2 **8** 1350 N/m^2 **9** $A = 47.5$ kN/m^2, $B = 18.1$ kN/m^2
10 (a) 101.5 kN/m^2, 11.5 kN/m^2; (b) 102.9 kN/m^2 **11** 31.1 kN/m^2,
28.9 kN/m^2 **12** 1.44 m   **13** 0.48 m   **14** 37 kN/m^2, 20 kN/m^2.

Index